Debugging Embedded Microprocessor Systems

Stuart R. Ball

Newnes

Boston   Oxford   Johannesburg   Melbourne   New Delhi   Singapore

Newnes is an imprint of Butterworth–Heinemann

 A member of the Reed Elsevier group

 Recognizing the importance of preserving what has been written, Butterworth–Heinemann prints its books on acid-free paper whenever possible.

 Butterworth–Heinemann supports the efforts of American Forests and the Global ReLeaf program in its campaign for the betterment of trees, forests, and our environment.

Library of Congress Cataloging-in-Publication Data

Ball, Stuart R.
    Debugging embedded microprocessor systems / Stuart R. Ball.
        p.   cm.
    Includes index.
    ISBN 0-7506-9990-6 (alk. paper)
    1. Embedded computer systems. 2. Debugging in computer science.
3. Microprocessors. I. Title.
TK7895.E42B34 1998
004.16—dc21   97-44582
                                                                                   CIP

British Library Cataloguing-in-Publication Data

A catalogue record for this book is available from the British Library.

The publisher offers special discounts on bulk orders of this book.
For information, please contact:
Manager of Special Sales
Butterworth–Heinemann
225 Wildwood Avenue
Woburn, MA 01801-2041
Tel: 781-904-2500
Fax: 781-904-2620

For information on all Butterworth–Heinemann books available, contact our World Wide Web home page at: http://www.bh.com

11 10 9 8 7 6 5 4 3 2

Printed in the United States of America

Contents

Introduction

This is a book about debugging. Solving problems. Making the thing work after you've built it and everything's gone wrong. Debugging seems to be fading as a skill among hardware and software designers alike, perhaps for the same reasons that fewer people today like to get under the hood of their car or build radios from scratch. Whatever the reason, in this book we look at the reasons why things go wrong in embedded systems, how to fix them, and how to avoid them altogether. We'll look at a few hypothetical situations, and at some that really happened.

For this book, we'll define an *embedded system* as one using a microcontroller or microprocessor, performing real-time, real-world functions. Examples of embedded systems are the microprocessor in a microwave oven or under the hood of a car, or the electronics that control a mail-sorting machine for the post office. Although a personal computer does not fall under this definition, we touch briefly on embedded applications using PCs as well.

The book emphasizes debugging techniques and tools, but more importantly, it considers ways to anticipate and prevent bugs. Bugs are easier to prevent than they are to find.

The audience for this book is embedded software and hardware engineers, engineers who are moving into the embedded field, and other engineers who want to understand embedded debugging. Engineers who have been designing real-time systems since Intel introduced the 8080 will get less out of the book than someone new to the field, but there is something in here for everyone.

This is not a book about digital design or software design, although we look at those issues as a means of avoiding debugging problems. I've assumed that you know the basics of microprocessor systems, such as the difference between address and data buses, what an EPROM and static RAM are, and enough about digital logic to know what an address decoder is.

1

Tools for Debugging Embedded Systems

Cody Horton stuck his head into the office. Matt Grover was on the telephone, but waved Cody to the empty chair beside his desk.

"Tell them to look at the parameters in the configuration table," Matt said to whomever was on the other end of the conversation. "I bet you'll find that one of them has been corrupted."

Cody looked around the office. A pen-and-ink caricature of Matt and someone, presumably his wife, was thumbtacked to the wall beside the desk. A stack of CDs sat in a plastic case beside the computer. He scanned the titles of the books in the bookcase, noting the usual assortment of IC databooks and application manuals. A red book with yellow lettering caught his eye, and he pulled it from the bookcase. *Morals of a Politician, Compassion of a Rattlesnake* was the title, followed by the subtitle: *Ethics in Today's Business Environment*.

Cody idly thumbed through the book before replacing it in the bookcase. He scanned the room again. A small circuit board was mounted on a wooden stand on one corner of Matt's desk. The board was fully populated with components except for an empty PLCC socket in the middle.

Matt hung up the telephone and Cody leaned across the desk and held out his hand.

"Cody Horton," he said. "I just started here a couple of months ago."

"I've seen you around," Matt replied. "What can I do for you?"

Cody pointed to the PC board on the stand. "What's that?" he asked.

Matt smiled. "I'll tell you the story about that sometime," he replied.

Cody shrugged and unrolled a schematic onto Matt's desk. "Are you familiar with the project we're doing for Quad Systems?" he asked.

Matt nodded. "I did some of the early estimates. Fifty test stations."

"One of the later additions to the contract specified that we provide hardware for in-circuit programming of the EPROMs on the boards. All the parts are soldered in the production versions so the machine vibration won't shake them out. Anyway, I got the task of designing the programmer. It's the first embedded design I've tackled on my own. Someone suggested that I let you look the design over before I start working on the prototype."

"Do you have a block diagram?" Matt asked.

In response, Cody pulled the last page from the stack of schematics and placed it on top. Matt studied the diagram for a few moments. "How do the boards plug into the programmer?" he asked.

"There's an adapter board for each board to be programmed. The adapter plugs into the 50-pin connector on the programmer and then into whatever connector is on the target board. Some of the boards were short of real estate, so we have to load addresses and data serially or something like that. We may use a short ribbon cable in some cases."

Matt took a marker out of his desk drawer and handed it to Cody. "Draw me a picture of the system," he said.

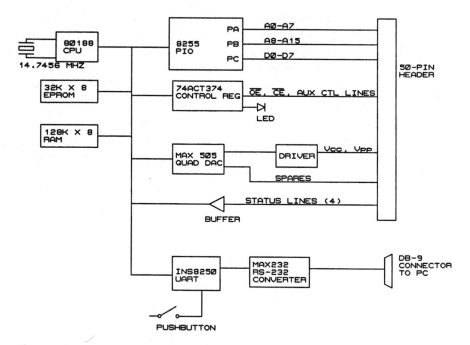

Figure 1.1 Programmer Block Diagram

Cody nodded and quickly sketched out a block diagram.

"What kind of PC?" Matt asked.

"The customer supplies the computer, so I'm trying to make it run on about anything. That's why I'm limiting the baud rate to 9600. That should let us use anything from a 286 to a Pentium."

Matt flipped through the schematic pages to the block diagram of the programmer, then took a calculator out of his desk drawer and punched some buttons. "I assume you picked the 14.75 crystal because it divides down to a standard baud rate. Did you make sure that the 16550 can handle the 7.37 MHz from the processor?"

"The UART input clock can go to 8 MHz."

"What kind of data gets passed over the serial interface?"

"The host computer sends commands and hex data to the programmer, and the programmer sends messages back to the host. The command interface hasn't been defined. I had planned to let whoever does the PC software supply a user-friendly interface, so the interface in the programmer could be kept fairly simple."

"Who is writing the firmware?"

"There was supposed to be someone from the software group assigned for the firmware and the PC software, but Josh Underwood

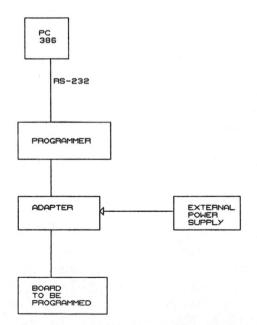

Figure 1.2 System Block Diagram

said they were pulled off to solve some crisis on another project. I'll have to do the firmware, but someone else will do the PC software."

"Have you added anything for debugging?"

"Like what?"

"Test points, diagnostic outputs, anything like that."

"I hadn't really thought about debugging the board."

Matt nodded again. "So you haven't thought about what tools you'll need to debug the design then."

"No."

"Do you know what tools we have in the lab?"

"I've used a 'scope. Other than that, I don't know what we have."

"Do you have a budget to buy any tools with?"

Cody looked sheepish. "No. This entire project was tacked on as kind of an afterthought, so there isn't much of a budget. I've been concentrating on the design itself."

Matt stood up. "Let's take a stroll out to the lab and see what's there."

Test Equipment

While Matt and Cody take an inventory of test equipment in their lab, we'll look at the types of test equipment that are used in debugging embedded systems.

Oscilloscope: Most engineers are familiar with the 'scope. It displays a horizontal trace of one or more waveforms, usually in real time.

DSO: A Digital Storage Oscilloscope (DSO, or sometimes Digitizing Oscilloscope) captures the data and stores it in an internal memory. This allows the data to be captured and examined later. A DSO allows infrequent events, such as a pulse that occurs only once every few minutes, to be displayed. An *analog 'scope* is better at displaying repetitive events than at displaying single events—a single sample of a waveform sweeps across the screen and is gone. A *storage 'scope* is a special kind of analog 'scope and uses a screen with a long persistence to "freeze" events, but these scopes have been almost universally replaced by the DSO.

A DSO passes the input waveform through an ADC (analog-to-digital converter), which digitizes the signal and then stores the resulting value in memory. Samples are taken at regular intervals, and the signal is reconstructed from the stored data. This method of capturing the waveform gives the DSO some unique limitations.

The first drawback to the DSO is limited memory. If a DSO can hold 1024 samples, then it can store and display 1.024 seconds' worth of data when sampling at 1 kHz, or 1.024 milliseconds' worth of data when sampling at 1 MHz. Obviously, the larger the memory, the more you can see at a given resolution.

DSOs also have a limitation in reconstructing the waveform. Figure 1.3A shows a waveform as it might be seen on an analog 'scope. Figure 1.3B shows the same waveform with dots where a DSO might capture the data. Figure 1.3C shows how this would be displayed on the DSO screen if the DSO connects the sample points with straight lines. Obviously, there is some distortion in the waveform. A higher sampling rate would improve the reconstructed image, but would also reduce the total sample time that can be stored. It is important to note that the voltages at the indicated sampling points in Figure 1.3B are all that the DSO stores—the waveform between these sample points is filled in by the DSO on the display only.

Some DSOs can use *equivalent time sampling* on repetitive waveforms. With this method, each trigger of the DSO causes samples to be

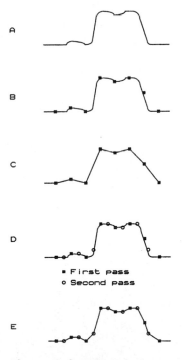

Figure 1.3 DSO Sampling

taken at the same regular intervals, but each sampling pass starts at a slightly different time with respect to the trigger. Figure 1.3D shows a waveform sampled twice, with the two sample intervals offset by half the sampling time. Figure 1.3E shows the resulting display, which is very close to the original signal in Figure 1.3A. Obviously, the more samples that are taken, the more accurate the reconstructed image will be. A nonrepetitive waveform or a single event cannot be sampled this way, as the DSO gets only one pass at storing the data.

Some DSOs improve the image by reconstructing the display with some type of curve match, such as a sine. This improves the appearance of the display, but it can make the waveforms appear to be better than they really are. Most DSOs that provide this feature also allow straight-line matching.

Aliasing, as shown in Figure 1.4, occurs when the DSO sample rate is too slow to represent the input waveform. When the sweep speed of an analog 'scope is set too slow, the screen will be filled with a signal that contains frequencies obviously too high to examine in detail. A DSO, however, does not capture or display anything that occurs between its regular samples. Figure 1.4A shows a repetitive waveform, such as the clock from a microprocessor. Figure 1.4B shows where a hypothetical DSO might sample the waveform, and Figure 1.4C shows the resulting display. Aliasing can be a real problem, especially if the

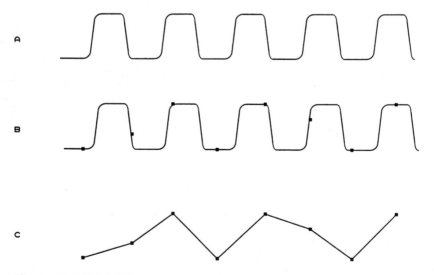

Figure 1.4 DSO Aliasing

sample time is very close to a submultiple of the input frequency. When that happens, it may appear that the waveform is correct, but at a frequency that is completely wrong. Anyone who has used a DSO for very long has been bitten by aliasing at least once.

Repetition Rate: Because of the time required to display the signal, a DSO typically cannot retrigger as quickly as a good analog 'scope. However, faster processors and memory are closing this gap.

Advantages of DSOs: With all its drawbacks, the DSO is still the instrument of choice for many applications, offering numerous advantages over the analog 'scope.

Hold: The first and most obvious feature of the DSO, and inherent in its sample/store method of operation, is its ability to hold data and display it indefinitely. This is particularly useful in debugging embedded systems, where an error may occur only once every few seconds or even minutes.

Repetitive Trace: A DSO can be used in a repetitive mode, where each new trace, instead of replacing the previous one, paints over the old data on the display. Figure 1.5 shows how the repetitive trace feature might be used to display the variation in the width of a pulse over several repetitions. The DSO triggers on the leading edge of the pulse, making the variation in this line very small. Each new trace paints over the old one, showing the variation in the trailing edge. In Figure 1.5,

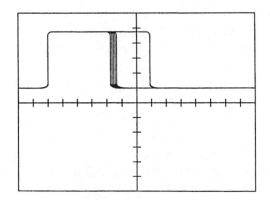

Figure 1.5 DSO Repetitive Trace

you can see that the trailing edge of successive traces are mostly grouped together, with one long pulse occurring (trailing edge past the centerline of the display).

Math Functions: Many DSOs provide the capability to perform mathematical operations on the stored data, such as integration, differentiation, or a Fast Fourrier Transform (FFT). While these features usually have limited application in debugging embedded systems, there are some cases where they can simplify a job.

Complex Triggering: The traditional analog 'scope is limited to triggering on a rising or falling edge, or on an external signal. Many DSOs, in addition to these capabilities, can trigger on runt pulses (where the peak falls between valid logic levels), narrow pulses, specific logic combinations on the inputs, and other more complex conditions.

Disk/HDD/Print: Many DSOs provide a floppy drive or internal hard disk, permitting configuration information and captured data to be stored on disk, where it can be examined and manipulated on a PC. A printer port is also a common feature, allowing the captured waveform to be printed.

Logic Analyzer: A logic analyzer samples a digital waveform and displays it numerically or graphically. A typical logic analyzer can store 8k (8192) samples, where each sample can be 16 to 160 bits wide. When sampling a digital waveform, a logic analyzer acts as a 'scope with a huge number of channels. There is one key difference between a logic analyzer and a 'scope: the 'scope shows analog signals, where the logic analyzer shows just a one or a zero. If the signal is higher than a threshold (typically 2–3v for CMOS), the logic analyzer shows it as a one. Otherwise, the signal is displayed as a zero.

Logic analyzer operation comes in two basic flavors: state and timing. In *state mode*, the logic analyzer captures data when an external clock (usually from the system being monitored) changes state. The clock may be enabled or disabled with *qualifiers*, and data may be captured on either the rising or falling edge. The key feature of state mode is the external clock. In a microprocessor system, the clock is usually connected to the system clock, or to the processor read and/or write lines, or to an address-ready signal (–AS on 68000-type processors, ALE on Intel-type processors).

In *timing mode*, the logic analyzer uses an internal clock to capture the data. Timing mode analyzers can be further subdivided into continuous and transitional analyzers. A *continuous mode analyzer* captures data at regular intervals. The sample interval is programmable from, say, every 10 ns (100 MHz) to every 100 ms (10 Hz). The total number of samples stored is the same for all sampling rates.

A *transitional mode analyzer* typically samples at its highest rate (100 MHz in the above example), but it stores data only when one of the input signals changes. The transitional mode analyzer also runs an internal timer at the sampling rate and stores the time with the stored data. When the information is displayed, the analyzer reconstructs the original waveform from the state changes and the time between them.

Transitional timing is the most common method of capturing timing information in modern logic analyzers. It greatly increases the amount of information captured, since there is no need to store samples between changes. Storing only state changes is the greatest advantage of the transitional analyzer, and also its greatest disadvantage. With a transitional analyzer, every state change is captured, even the ones you don't want—such as when the address bus is changing between cycles. Every state change takes up more storage. A continuous sampling analyzer would ignore these changes, capturing changes only on its regular clock.

State analyzers and both types of timing analyzers typically allow several signals to be grouped and displayed as a bus. Most modern analyzers can operate in either state or timing mode, or both.

Figure 1.6 shows a generic trace from a logic analyzer in state mode and in timing mode. In Figure 1.6A, a hypothetical processor executes instructions from memory. The 16 address lines are grouped and displayed as a single 4-digit hex word. The 8 data lines are grouped and displayed as a single 2-digit hex word. The time that each sample was stored is displayed as well. Usually, the time can be absolute, as shown, or relative, showing only the time between successive samples. Not all analyzers have the capability to display the time. Such capability usually exists in analyzers that have transitional timing mode. To get the display shown in this example, the analyzer would typically be clocked with the –WR and –RD lines (for Intel-type processors) or –DS (for Motorola-type processors). Note that the trace starts with state numbered –0004. This indicates that the trigger occurred at state 0000, where the address is 0024. Negative state values typically indicate samples that were stored prior to the trigger.

Figure 1.6B shows the same processor cycles as they would look in timing mode. Note that the ALE signal (which latches the address on Intel-type processors) is a single signal and is shown as such. The address and data buses are still grouped as 16-bit and 8-bit words respectively.

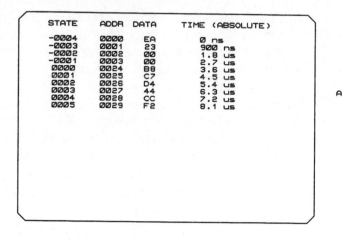

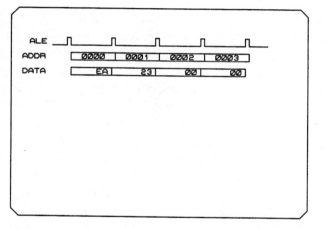

Figure 1.6 Hypothetical Logic Analyzer Traces, State and Timing Modes.

The displays in Figure 1.6 are for a nonexistent, hypothetical analyzer. A real analyzer typically indicates which sample was the trigger point; may have a scale at the top of the screen in timing mode, marked with regular time intervals; and would usually have two or more cursors that can be used to measure the time (or number of states) between two points. Other features of a real analyzer might include the ability to display state data in timing format and vice versa; the ability to display data in ASCII, binary, octal, decimal, or other formats (instead of hex); and numerous other features.

When clocking a state mode analyzer, it is often necessary to capture data when one of several signals goes active. For example, Intel-type processors have separate read and write strobes. Most state analyzers permit more than one input clock and will capture data on any clock that is enabled by the user.

The channels in a logic analyzer are usually grouped into pods, each of which contains 8 to 16 channels. Many analyzers provide two internal "machines" that can be used simultaneously. This feature typically allows the user to group some pods to one virtual machine and other pods to the other machine. Each "machine" functions as a logic analyzer, either independently of the other or cross-connected with each other. Some analyzers require that, if two virtual machines are used, one be in state mode and the other in timing mode. Other, more powerful, analyzers permit both machines to be state or timing mode. The most powerful of these allow data captured on the two machines to be displayed side-by-side and time-correlated. This is a great help in debugging multiple-processor systems. Figure 1.7 shows how data from two separate systems might be displayed on our hypothetical logic analyzer. Data captured for virtual machine 2 is underlined in the figure. In a real analyzer, data for the second machine might be underlined, highlighted, or shown in reverse video.

The first column shown in Figure 1.7, STATE, shows the state number for both machines. As can be seen, the two processes do not need to be synchronized, as machine 1 is at state 0000 while machine 2

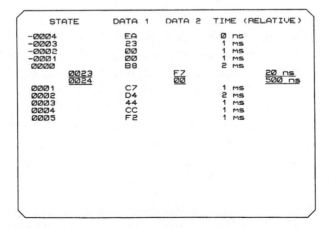

Figure 1.7 Logic Analyzer Display Showing Two Time-Correlated Virtual Machines

is at state 0023. The two data points captured for machine 2 occurred between states 0 and 1 for machine 1.

Many logic analyzers permit one virtual machine to capture data in state mode and the other to capture data in timing mode. The better analyzers permit the data to be correlated, either directly on the display, or by correlating the position of reference cursors. All analyzers that support two virtual machines permit one machine to be triggered by the other. This allows you to trigger one machine when an error occurs and see what the other machine was doing at that time.

Integrated DSO: Some logic analyzers, such as the Hewlett-Packard HP1660 series, have an option that can add two or more high-speed DSO channels. The DSO can trigger the analyzer or be triggered by the analyzer, but the best feature is that the DSO and analyzer data can be time-correlated. This is a powerful combination that greatly simplifies debugging real-world problems. The DSO could trigger the analyzer when, say, a processor-controlled voltage rises above a predetermined level. The analyzer could then be used to determine where in the code the processor was executing when the event occurred.

Disassembler Pods: Many logic analyzers have optional disassembler pods. This is usually a pod that plugs into the target processor socket and software that configures the analyzer and allows it to disassemble the executed code, showing assembler mnemonics on the screen.

Glitches: One problem that can occur in any digital design is a "glitch," which is a very short pulse, usually caused by a race condition. Newer analyzers can trigger on a glitch and usually display glitches in a different color, highlighted, or with a special symbol. The definition of a glitch varies from one manufacturer to the next, but is usually something between the smallest pulse the analyzer can detect and the shortest sampling interval. Some older analyzers with continuous sampling defined a glitch as a pulse shorter than the sampling interval, whatever that was set to by the user. This was a feature that was extremely useful in debugging certain interrupt problems, but most of the newer transitional-mode analyzers have gone to the fixed glitch definition instead, making this capability unavailable.

Triggering: Triggering capabilities of logic analyzers are extensive and vary greatly from manufacturer to manufacturer. It would be impossible to catalog them all here, but certain trigger features are both fairly common and useful in debugging embedded systems.

State/Timing Word: All analyzers permit triggering when a particular pattern is detected on the state and/or timing lines. Timing words only have to occur long enough for the analyzer to detect the condition. State words have to be stable when the data is captured by the external clock. Some analyzers have programmable filters so that a timing word must remain stable for some time before it will be recognized. Equivalent state filters require the word to be active for a certain number of states before the trigger is allowed.

If/Then/Else: Most analyzers also have some kind of conditional trigger, something like If state word A occurs, followed by state word B, then trigger; but if state word C occurs, go back to looking for state word A. These can often be nested several levels deep and permit loops back to higher levels.

Conditional Storage: This feature permits storage to be turned off and on to save memory. For example, you might turn storage off when the processor is in the address range where the interrupt service routines (ISRs) are stored, so that a regular timer tick interrupt doesn't fill the storage buffer.

Range Checking: This allows the analyzer to trigger when a value is above or below a particular value (say, if the value on the address bus exceeds 2000h). Range checking can also permit triggering if a value is within or outside a specific range, such as if the address is between the start and end addresses of a particular ISR, or if the address goes above or below the valid code space.

The logic analyzer figures included so far have been simplified for clarity. Figure 1.8 shows an actual transitional timing trace of an 8031 microprocessor, captured on a Hewlett-Packard 1660-series logic analyzer. This trace was captured and stored to disk as a black-and-white .TIF file. Note that the actual trace has additional information left off of the previous, simplified examples. The display can be zoomed in or out to look at the data, and the machine has time markers (not enabled in this trace) that can be used to determine time differences between different points on the display. The address and data lines are grouped as buses on the trace, but they can also be displayed as individual signals.

Emulator: A microprocessor emulator replaces the microprocessor chip, plugging into the socket that the processor would otherwise occupy. The emulator then operates exactly the same way as the processor (ideally), executing code just as the processor would. The emulator adds the

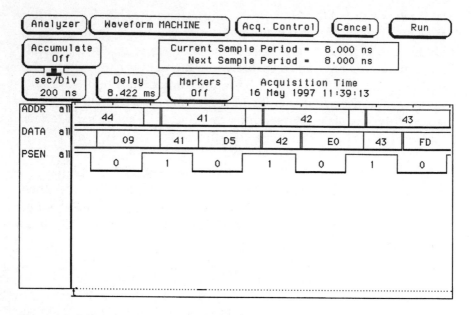

Figure 1.8 HP 1660 Transitional Trace

capability of stopping execution and allowing you to examine the contents of memory, internal processor registers, and other characteristics. Photo 1.1 shows a UEM-series emulator from Softaid. The host computer, which provides the user interface, would connect via a cable to the box in the photo, and the circuit at the end of the ribbon cable would plug into the socket of the target system.

Most modern emulators are connected to a PC. The PC provides the user interface, usually using Windows. Figure 1.9 shows the sort of information that might be displayed on a generic Windows-based emulator screen. The screen is divided into windows, each of which can usually be sized and moved to different locations. The window showing code in the figure (8031 code in this case) displays what the processor was doing when emulation was stopped. The trace buffer, at the bottom, would show complete machine cycles. The difference between the two is that the trace buffer shows actual read and write cycles, with timing information, while the code buffer just shows what code is at a particular address. The source window would show the source file, which may be in assembler or an HLL. Typically, the source file will scroll so that it always shows the source code corresponding to the code displayed in the code window. When the source file is in

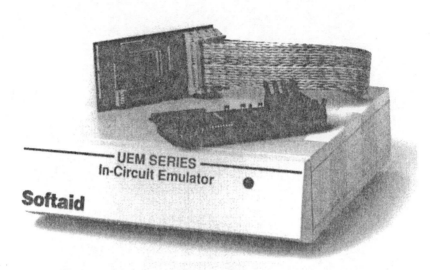

Photo 1.1 Softaid UEM-Series Emulator (Courtesy Softaid)

assembler, of course, the two will be the same. The register window shows contents of internal registers.

The toolbar would have Windows buttons that can be activated by using the mouse. Typical buttons might be:

Resume: Causes execution to resume at the point where it was stopped.

Break: Allows a new breakpoint address or condition to be set.

Reset: Resets the processor, causing new execution to begin at the reset address.

Step: Single steps the processor, executing the next instruction in sequence.

Not all emulators, of course, will have exactly the same features. An 8052 emulator, for example, might have a window dedicated to showing the contents of special function registers. An 80188 emulator might have a window that shows the contents of the internal Peripheral Control Block.

Breakpoints: An emulator is normally used by running the code until an error occurs. You then set breakpoints in the code to determine what

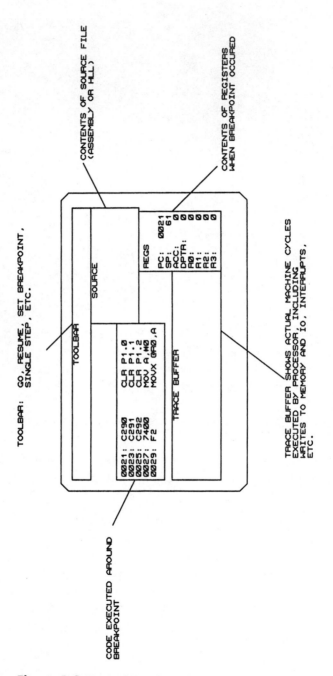

Figure 1.9 Typical Emulator Screen

went wrong. A breakpoint typically stops the processor when it executes code at a particular address. The emulator then captures and displays the contents of the various CPU registers. For example, you might have a problem in some code where the processor gets the wrong value from a table. You set a breakpoint at the place in the code where the error is first detected. When the emulator stops, you look at the memory to see what was wrong with the pointer variable.

Breakpoints are typically set at an execution address, but many emulators offer other breakpoint options. For example, you might be able to set a breakpoint when a particular memory location is written with a specific value. Some emulators have external trigger inputs that let you stop execution with the trigger output of a logic analyzer or DSO. This lets you see what code was executing when a particular hardware event occurred.

Depending on its sophistication, the emulator may provide trace capability. In the pointer example mentioned above, suppose that looking at the pointer variable doesn't give you any clues to how the variable got set to the wrong value. In this case, you might use the emulator to trace the variable. In this mode, the emulator will keep track of every read and write to that address, so you can look at the trace buffer to see what went wrong and when.

The emulator trace buffer can usually be turned on and off, much like the conditional storage in a logic analyzer. This lets you trace, for instance, only the machine cycles that happen inside a particular ISR.

Emulators that support *source level debugging* are designed to work with your compiler and let you set breakpoints at the source level. If you are debugging in C, you could set a breakpoint at the label ERROR_SERVICE, and the emulator software would take care of figuring out which assembly/machine instructions that label corresponded to. When the program stopped, you could look at the variable ERROR_CODE by name without having to look at a link map to see what address the linker assigned to it. Source-level debugging depends on the emulator getting a variable map from the compiler, so it knows where everything maps to in memory. The emulator has to understand the map, which means you have to pick an emulator that is compatible with your compiler (or vice versa), or you have to convert the compiler-generated map to a format the emulator understands. This is not as trivial as it sounds; if you are working in C, say, and you want to examine a variable by the name it was given in the source code (as opposed to its absolute address), the emulator must know what kind of variable you are looking at so it will know how to display it. The emulator must be able to distinguish integers from floating-point variables, for example.

Emulators come in various flavors, depending on how much you have to spend and on the type of processor supported. Some simple emulators, for example, are basically ROM programs that get control and do all the emulator functions in software. Emulators such as these cannot emulate your code unless you run out of RAM, because breakpoints are set by replacing instructions in RAM with some type of software interrupt that transfers control to the emulator program.

Microcontroller chips, with internal ROM and RAM, are a particular problem for emulator manufacturers. Since the code in these chips is executed internally, an external ROM program is useless. Some lower-cost emulators force you to use the microcontroller chip in external memory mode so they can use a ROM program. If you are using a microcontroller and want to run out of internal ROM, you will need an emulator that uses a *bond-out chip*. This is a chip from the IC manufacturer that has enough internal nodes brought out of the package to allow an emulator to get control. The key thing to remember in buying a microcontroller emulator is to be sure the emulator doesn't use chip resources that you need for your application.

Single Stepping: Single stepping involves executing the program one instruction at a time. The emulator does this by, effectively, setting a breakpoint after each instruction. Each time you step, the breakpoint is moved out one instruction. Single stepping is a very good way to trace the logic of program flow. The problem with single stepping is that external events, such as motor shaft encoder pulses, interrupts, and so on, just keep coming while the emulator is stopped. Single stepping has limited use in these cases, unless the external events can be turned off. In an emulator that supports source-level debugging, you can usually step through the high-level code without having to step through each assembly instruction that makes up an HLL statement.

Timers: Many emulators provide timers that can be used to determine how long an instruction takes to execute, or how long a group of instructions (such as an ISR) takes to execute.

Memory: Most emulators provide RAM memory. This can usually be mapped so that some of it emulates ROM (the target processor cannot write to it) and the rest of it emulates RAM. It is common during development to run the program out of emulator RAM, never programming an EPROM until the software is completely debugged. Some emulators have more flexible breakpoint capability when running out of RAM than when running out of ROM.

ROM Emulator: Figure 1.10 shows a block diagram of a ROM emulator connected to a target system. Figure 1.11 shows the block diagram of a simple ROM emulator. Data received from a host PC is stored into RAM. Each time a new byte is stored, the address counter is incremented, causing the next byte to go in the next successive location. When the entire program is loaded into the ROM emulator RAM, the address multiplexer is switched so that the target processor provides the address and gets the data from the RAM. The ROM emulator now looks like a ROM to the target system.

ROM emulators are primarily used to speed up the development process. Instead of programming EPROMs, and always having to keep some EPROMs erased, the code is downloaded into the ROM emulator and executed there. The ROM emulator can usually be loaded much faster than a PROM can be programmed, so it saves time. Some ROM emulators provide an output to hold the target system in reset during download.

Debugger: A debugger, or software monitor, is a program that resides in memory with the application program and that provides some of the same features as an emulator. Depending on its degree of sophistication, a debugger can examine and alter memory contents, download code from a PC host, single step through the code, and set breakpoints. To set breakpoints, of course, the program must be run out of RAM. Since the debugger program is dependent on a breakpoint to get control, it cannot monitor writes to particular addresses, stop on an external

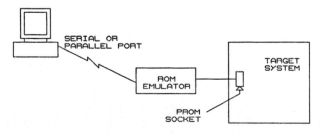

Figure 1.10 ROM Emulator Block Diagram

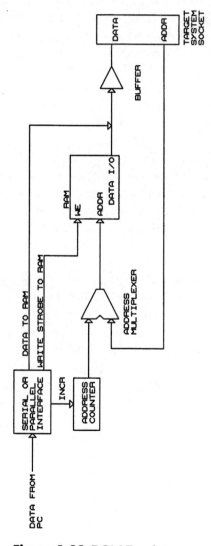

Figure 1.11 ROM Emulator

event, or provide many of the advanced debugging features of sophisticated emulators.

Some off-the-shelf processor boards come with a debugger installed, and some engineers like to embed a simple debugger in the ROM with the application code. When a simple debugger is embedded in the ROM, the reset code usually branches to the debugger program when a particular shunt jumper is installed, or a DIP switch is set, or a similar event occurs.

A debugger usually sets a breakpoint in the code by replacing an instruction with the breakpoint instruction. For example, on the 8086 family of processors, the INT 3 instruction, with opcode CC (hex), has the same effect as a hardware interrupt, causing the processor to service an ISR. The software monitor would replace the opcode in (RAM) memory with this opcode, and when this instruction was executed, the debugger program would get control and display register contents, or whatever else needed to be done. To move the breakpoint, the original opcode would be put back in its proper place, and the breakpoint opcode would be put in the new break location. When the software executes an instruction that has essentially the same effect as an interrupt, it is sometimes referred to as a *software interrupt*. Some processors implement a single instruction just to make debugging easier (this is the purpose of the x86 family INT 3). Of course, other processors have different software interrupt opcodes; in the 8085/Z80 families, for example, RST 7 (0FF hex) forces the processor to address 38h.

To use a debugger, you have to provide a serial port for communication with the PC, and you have to reserve a single-byte software interrupt (even if it is the only one that the processor supports) for the monitor to use. To use breakpoints, the system must have enough additional RAM to hold the code. The drawback to using a software monitor, besides the system resources it requires, is that the debugger is susceptible to catastrophic failures in the code. For example, if the code writes all through memory, destroying the stack, or changes the UART baud rate, the debugger may be unable to function or may give unreliable results.

The user interface of most debuggers is simpler than that of an emulator. Since the debugger in a simple ROM-based system does not have a cross-reference table in memory, you cannot perform source-level debugging. However, a debugger can connect to a PC that provides a user interface and can contain the cross-reference table and other information needed for source-level debugging.

Photo 1.2 shows a screen capture of the VisualProbe debugger from SSI. This screen contains windows displaying memory contents, the source code, and CPU registers. In a typical Windows-based debugger,

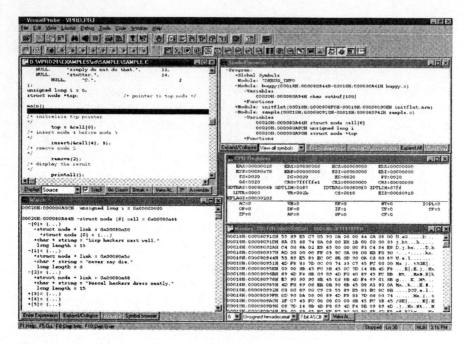

Photo 1.2 VisualProbe Debugger (SSI, Irving, CA)

you can size these windows, eliminate them, or replace them with windows containing other information.

On-Chip Debugging Resources: As processor clock rates increase, plug-in emulators become more difficult to build. Even a short cable connecting the emulator to the target socket introduces unacceptable skew in the signals. It is difficult to make a multichip emulator design run as fast as the target CPU, where everything is on the chip. To make debugging easier, many of the more complex processor designs include some debugging capability on the CPU chip itself. You typically see this capability in processors of about the complexity of the 386 and up.

As an example, let's look at the debugging resources in the PC-type processors such as the 386, 486, and Pentium. The x86 family of processors includes a special single-byte instruction for debugging, INT 3, which has already been described. The TI data book for the 486 shows that this chip has six debug registers, as shown in Figure 1.12. Four 32-bit registers (called DR0–DR3) provide breakpoint addresses. Another register, DR6, provides status. DR7 is used to determine the type of breakpoint. Breakpoints may be set on instruction fetch, writes

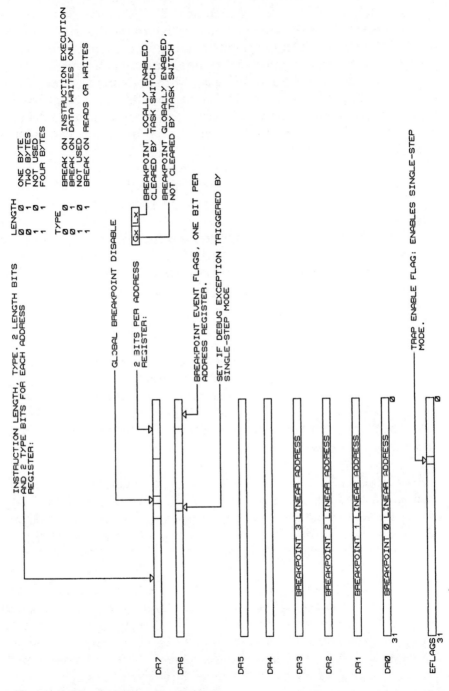

Figure 1.12 x86 Debugging Registers

only, or read/write (there is no read-only breakpoint). There are two bits for this purpose in DR7 for each of DR0–DR3. In addition to the type of breakpoint, there are two bits that determine the word size for each of the breakpoints. Other bits control what happens to breakpoints under various circumstances when a task switch occurs.

In operation, you would write an address to one or more of the breakpoint registers, write the DR7 register with the type of breakpoint (or breakpoints), and then execute the program you are debugging. When the breakpoint is reached, the hardware causes the processor to, in effect, execute an INT 3.

In addition to the breakpoint registers, the TI 486 also has a bit (called TF) that can be set in an internal flag register (EFLAGS), and that will cause a breakpoint after every instruction: in effect single-stepping the program.

Other advanced processors in the x86 family have similar capabilities; a random sampling of x86 databooks reveals that the Intel 386EX (a version of the 386 optimized for embedded applications), the AMD 386DE, and the SGS-Thompson 6x86 have debugging registers identical to those of the TI 486.

Background Debugging Mode: Motorola took a slightly different approach to debugging support with some of their processors. The 68332, for example, provides for a 17-bit synchronous serial interface that permits debug commands to be issued to the processor. The Motorola scheme typically doesn't include breakpoint registers (although the Power PC does), but it does let you examine and change memory and registers, and it lets you single-step the code.

JTAG Port: JTAG is a serial interface that was originally intended for testing a complex IC, such as a large ASIC or microprocessor. JTAG allows the state of the IC pins to be loaded into a register and read serially. Some manufacturers have started to provide access to their internal debug features via the JTAG port.

Using On-Chip Debug: In general, you won't use the on-chip debug features directly. In most cases, you will purchase an off-the-shelf debugger program that takes advantage of the CPU's debugging hardware and provides a friendly user interface. Using the on-chip hardware provides a debug capability somewhere between that of a software monitor program and that of a full emulator. Unlike the software monitor, with on-chip hardware you can set breakpoints or single-step without having to run out of RAM (a major advantage in ROM-based embedded systems).

If the debug hardware supports it, you can set breakpoints when a memory location is read or written.

On the downside of using on-chip hardware, you will give some things up. First, the hardware breakpoint capability is limited, although additional software breakpoints may be provided. The TI 486, as already mentioned, will permit breakpoints on memory read/writes. Other processors may have only instruction breakpoints—they may not be able to monitor memory accesses. Even the TI 486 can break only when a particular location is read or written—it cannot monitor a specific location for a particular data value in the way that some emulators can. On-chip debug resources are limited; the TI 486 allows for four simultaneous breakpoint addresses, and some emulators support many more. Last, but possibly the most important, you don't get any real-time trace capability with on-chip debugging resources.

This discussion about the tradeoffs between on-chip debug support and full emulation is somewhat misleading. Because of the problems with emulating high-speed processors, your choice is usually not between on-chip debugging and an emulator. The choice is usually between on-chip debugging and nothing, at least when the processor chip first comes out. Internal debug capabilities are probably destined to be the standard debugging method for complex processors. This is something of a drawback, as the on-chip capabilities are usually much less extensive than those provided by a full emulator. The difficulty of building a full-speed emulator for today's complex microprocessors would appear to be pushing the industry toward less debugging capability even as the designs become more complicated.

Simulator: Simulator software lets you run your application on another computer, usually a PC or workstation. Simulators are useful for testing and debugging code before the actual hardware is available, and they also allow multiple software developers to work in spite of limited hardware availability. In addition, a simulator—such as the Visual-Probe simulator from SSI—provides a crash-free environment: if the software crashes, the simulator doesn't, which permits you to see what happened. Photo 1.3 shows a screen shot of the Visual Probe simulator software.

PROM Development Tools: These tools are the basic PROM programmer and eraser. Although they may not seem to be directly related to debugging, they can be a real bottleneck. If you are not using a ROM emulator, so you have to program PROMs during development, you will want a PROM programmer and eraser located near where you are doing the

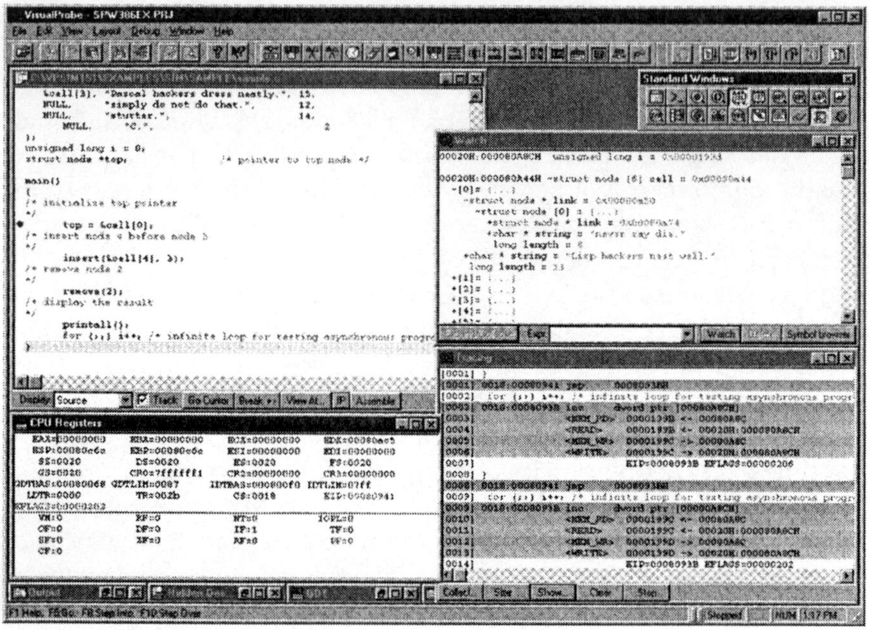

Photo 1.3 VisualProbe Simulator (SSI, Irving, CA)

debug. If you have to walk down the hall and into another lab to erase your PROMs, you will tend to let them lie around the bench until you need to make a change and discover that none of your PROMs are erased. Then you've got a 20-minute wait before you can proceed.

Flash PROM and other EEPROM (Electrically Erasable PROM) technology makes erasing easier—the programmer can erase the parts before programming them, making a UV eraser unnecessary.

Some PROMs and microcontrollers come in a OTP (One Time Programmable) version. These parts contain an EPROM, but without the erasure window, so they can be programmed only once. If you are using a microcontroller with internal PROM, and you plan to use an OTP version in production, make sure you have plenty of the EPROM versions of the chip for development. If the OTP version has slightly different pinouts or differs in other ways from the EPROM version, make sure you have a version of the board modified to let you debug with the EPROM version—otherwise you will throw away a lot of the OTP parts debugging the code.

Counter/LED Circuit: Figure 1.13 shows a simplified version of a circuit that I like to keep in my toolbox. It's a counter driving LEDs, one LED

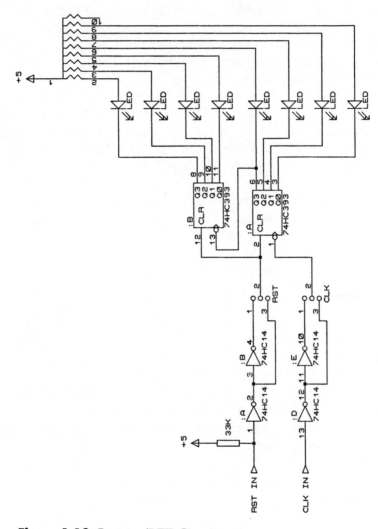

Figure 1.13 Counter/LED Circuit

per counter bit. A 74HC14 allows the counter to increment on either the rising or falling edge of a signal, and the counter can be reset when another signal is either high or low. I like this circuit because it offers most of the advantages of a logic probe, but with several enhancements. A logic probe will tell you whether a signal is pulsing, but not how fast (unless the pulse rate is really slow). But by looking at the LEDs on the counter circuit, you can tell approximately how fast the signal is. If multiple pulses occur close together, a logic probe will blink once; the counter circuit will actually show you how many pulses occurred.

The schematic as shown will display a "zero" bit as having the LED on. This is the reverse of what you normally expect. When I actually build these, I usually insert an inverter (such as a 74HC240) between the counter and the LEDs so that a "one" lights the corresponding LED. I also like to build 16 or more bits for more versatility, and I often put the entire circuit in a PLD or FPGA. But any way you build it, this is a useful tool to have around.

The Tool Vendor: Although not usually considered a debugging tool, the technical support provided by your compiler, kernel, or emulator vendor can be very important. When you have spent days on a bug that has no apparent cause or when the code just doesn't seem to be doing what you wrote, then you might want to call the tool vendors. Sometimes such a call will result in nothing but finger-pointing, but sometimes you will get an answer that goes like this: "That function has a bug. Code around it this way, and we'll fix it in the next update." If you are designing a complex system for which you are purchasing a new compiler, kernel, emulator, and other tools, you might want to look at how your vendors handle technical support. You might consider purchasing two or more components (kernel and debugger, for example) from the same vendor to avoid one layer of finger-pointing when things don't work.

This chapter has provided an overview of the basic tools used in debugging embedded systems. We've left out the most important tool, which is of course what you get paid for: your ability to analyze the problem. In the chapters to come we look at some ways this and the other tools described here can be put to use.

2

Adding Diagnostic Helps to the Target System

For the second time in as many days, Cody stepped into Matt's office. Matt turned around from his computer, where he was entering a schematic.

"What did you find out?" Matt asked.

"We have an 80188 emulator," Cody replied, "but it's in use on another project, and that project is late. I doubt I could even get it after hours."

"How about logic analyzers?"

"We've got two, but they are both being used. I might be able to borrow one on a part-time basis, but both of the other projects are higher priority than the programmer."

"Until it's time to deliver the programmer and it's not ready."

Cody nodded. "I'd be tempted to change processors if the PC boards weren't nearly finished. It looks as if the only tool I can get is a 'scope. What do you suggest?"

Matt leaned back in his chair, and it squeaked loudly. "I'd say we need to look at your system to see what we can do to make debugging easier." He pointed to the system diagram that Cody had drawn the day before, which was still on the whiteboard. "What kind of data gets sent back and forth to the PC?" he asked.

Cody tapped the sketch with his finger. "I was planning to send single-byte opcodes to the PC for various things. You know, 01 to command a blank check, 02 to start programming, that sort of thing. Whoever does the PC software will make a Windows program that hides all that from the user."

"What goes back the other way?"

"Completion codes, error codes, simple stuff."

"List the commands the programmer needs."

Cody picked up a marker and made a list:

Program device
Blank check device
Verify device
Select device type
Fill buffer memory with constant
Set memory offset
Set device offset
Download Intel-format hex file
Download delimited/undelimited hex file
Download binary file
Select operation size

"Some of these commands obviously have data associated with them. I planned to send that as binary values, most-significant byte first."

Matt steepled his fingers and rested his elbows on the desk. "I'd scrap the opcode idea. Use ASCII codes instead. That way, you can debug the programmer using a PC and an off-the-shelf communications program. No need to wait until the software person becomes available before you verify the interface. Make the parameters you send ASCII also."

Cody tossed the marker back. "Show me what you mean." he replied.

Matt came around the desk. He thought a moment, then wrote beside the list Cody had made:

#P: Program device
#B: Blank check device
#V: Verify device
#T: Select device type
#F: Fill buffer memory with constant
#M: Set memory offset
#R: Set device offset
#DI: Download Intel-format hex file
#DH: Download delimited/undelimited hex file
#DB: Download binary file
#S: Select operation size

"Why the pound sign in front of the commands?" Cody asked.

"So you can differentiate them from, say, hex data."

"So now my parameters have to be converted to hex from ASCII."

"But look what you get in return. And you have the conversion code anyway, for the hex download."

"Okay. I see your point about debugging the code with a communication program. This looks good. What else?"

"You've got 128k of RAM on the schematic. How much will the program use?"

"I planned to leave the upper 64k as a buffer for programming data, and the lower 64k for the code to use. I don't expect the program to need more than a couple of thousand bytes, with stack and everything."

"I'd add a couple of commands then, to make debug easier."

"Like what?"

Matt added to the bottom of Cody's list:

#XB: Display 64 bytes of programming buffer at specified address

#XM: Display 64 bytes of memory at specified address

"This will let you examine variables in memory and also let you look at the contents of the programming buffer while you're debugging the download code," Matt said.

"Think that will do it?"

"I'd add one other thing. A trace buffer."

"What's a trace buffer?"

"It's a chunk of memory that you set aside to monitor system activity. The code writes values to the trace buffer when it executes significant points in the code. When something fails, you can look at the trace buffer to see what the history was. For example, you might put a value, a code, of 3 in the buffer when you start programming a device. 01 might mean a reset occurred. Maybe 05 when you decode a command from the host PC."

"What happens when the pointer gets to the end of the buffer?"

"You just wrap it around to the beginning. You keep a pointer in memory or a termination value in the buffer so you can find the end."

Cody nodded. "That seems easy enough. I assume you've coded this before?"

Matt took a worn vinyl notebook from the bookcase and flipped through it. "Here it is," he said. "Already coded for an 80188 processor."

Cody looked at the listing for a moment. "What if the buffer isn't a binary length?"

Listing 2.1 80188 Trace Buffer Code

```
; diagnostic code, writes byte in AL to diagnostic
; buffer, DIAGFIFO.
; DIAGNOSTIC will write a value of 0FFh after
; the requested byte is added to the buffer, to
; mark the end.
; The buffer wraps around, overwriting
; previous data. DIAGNOSTIC saves registers it
; uses (except AL), and disables interrupts.
; Note that DIAGNOSTIC turns interrupts back
; on at the end - don't call from within a routine that
; needs interrupts disabled, unless the CLI/STI is
; removed from DIAGNOSTIC.

; DIAGNOSTIC needs one variable in memory,
; DIAGPOINT, which maintains the current pointer.
; DIAGNOSTIC also must know the start address of
; DIAGFIFO, the trace buffer in RAM.

; DFIFOLEN is a binary value that is ANDED with
; the DIAGPOINT to force a wraparound
; at the end of the buffer. The buffer must be a
; binary length - 256 bytes, 512, etc. DFIFOLEN
; would be 0FF for a 256-byte buffer, 1FF for a
; 512-byte buffer, etc.

diagnostic:
    cli                         ; Disable the intr
    push si
    mov si,diagpoint            ; Current buffer pointer.
    mov diagfifo[si],al         ; Store the requested value.
    inc si                      ; Incr the pointer,
    and si,dfifolen             ; and force a wraparound.
    mov diagpoint,si            ; Store the new pointer.
    mov diagfifo[si],0ffh       ; Mark the end of the buffer.
    pop si
    sti                         ; Enable the intr again
    retn                        ; All done.
```

"Then where the code does an AND operation to force a wrap-around, you have to do a compare with the buffer length, and force the pointer to zero if it has reached the end of the buffer. It's more general that way, but takes a little longer to execute. I was interested in speed when I wrote this code, so I did it this way."

"Why write the terminator to the buffer? Can't you just use the pointer to find the end of the buffer?"

"Good thinking. The terminator byte just makes things easier, since you don't have to remember where the end is after you look it up. To use the trace buffer, I'd add another command to your interface protocol."

"That would be?"

Matt picked up the marker and wrote:

#XF: Clear diagnostic FIFO to zeros

"Why not just clear it at powerup?" Cody asked.

"Because if the software locks up, you might need to reset the hardware. In that case, you'll want a history of what went wrong."

Cody picked up the legal pad he'd brought with him. "Okay," he said. "I'll make the command protocol use ASCII characters and data, I'll add commands to look at memory and flush the trace buffer, and I'll add the trace buffer. Anything else?"

"Have you done a walkthrough of the code with anyone?"

"No. This design has been fairly informal."

"You should have. You can catch a lot of bugs that way. We'll skip it for now. You've got enough to keep you busy for a day or so."

While Cody modifies his code, we'll look at the trace buffer and some other things you can add to your design make debugging easier.

Trace Buffer

The trace buffer that Cody is adding to his software is a powerful technique for most embedded designs, if there is enough memory to support it. Typically, when something goes wrong in an embedded system, you want to know what happened just prior to the failure. The trace buffer already described will be a binary length (256, 512, 1k, etc.), and each location will hold 1 byte. Each byte is an action code that indicates some specific action. For example, let's take a hypothetical signal-monitoring system that monitors both a continuous stream of digital data passing between two other systems and commands over a serial port from a host PC. The system, shown in block form in Figure 2.1, has to analyze the data stream for any of three possible patterns of bytes. If any of the three is detected, the monitor sends a message to the PC. The

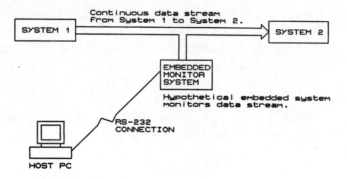

Figure 2.1 Hypothetical Embedded Data Monitor
System

PC can mask the patterns individually, and if a pattern is masked, the
monitor will detect it but not send the notification to the PC. Let's further say that the monitor has a regular 50-millisecond (20 Hz) interrupt
that it uses to measure timeouts and other things.

We could construct a table of action codes for this system that
might look like this:

01: Powerup reset.

02: Host PC signaled that it received a powerup reset.

03: Pattern 1 start detected.

04: Pattern 1 verified, but host PC had it masked.

05: Pattern 1 verified, notification sent to host PC.

06: Pattern 2 start detected.

07: Pattern 2 verified, but host PC had it masked.

08: Pattern 2 verified, notification sent to host PC.

09: Pattern 3 start detected.

10: Pattern 3 verified, but host PC had it masked.

11: Pattern 3 verified, notification sent to host PC.

12: Command to mask pattern 1 received from host PC.

13: Command to mask pattern 2 received from host PC.

14: Command to mask pattern 3 received from host PC.

15: Command to unmask pattern 1 received from host PC.

16: Command to unmask pattern 2 received from host PC.

17: Command to unmask pattern 3 received from host PC.

64: 20 ms interrupt service.

For this system, the last 32 bytes of a trace buffer might look like the following:

64 64 03 04 64 03 04 15 16 17 64 03 05 64 06 08

13 03 04 64 06 07 64 03 64 04 64 64 06 16 07 FF

If this was what we captured after an error occurred, we could analyze it like this: First, the FF value at the end of line 2 indicates the end of the buffer. Working backward from that, we can see that there was a sequence where the start of pattern 2 was detected, and a command was received from the host PC to unmask pattern 2, but when the end of pattern 2 was detected, the code still thought it was masked: (06 16 07 at the end of line 2, underlined). This might or might not be an error, depending on how the system was specified. Let's say that it is an error, and that the code to indicate the presence of pattern 2 should have been sent to the host PC. We might suspect an error in one of three places, based on this sequence:

1. The host PC might be too slow to unmask the pattern, and might be getting the command to the monitor system too late.
2. The firmware in our monitor system might fail to recognize an unmask command if it occurs during the pattern being unmasked.
3. If our monitor system is designed to handle unmask commands during the detection of a pattern, then the problem could be a race condition. The trace buffer gives the appearance that all commands were received in regularly spaced intervals. But it is possible that the last command to unmask pattern 2 occurred just immediately before the end of pattern 2 was detected. Suppose the PC commands are written to the trace buffer from an interrupt routine that processes the serial inputs. The unmask command might have set a bit somewhere to indicate that pattern 2 is unmasked, but the code might have already tested that bit and decided that the pattern was, in fact, masked. Because the interrupt could have occurred just prior to writing the 07 code to the trace buffer, this is a plausible explanation.

Expanding the Trace Buffer

Clearly, we might want to put more information in our trace buffer to track down this hypothetical problem. We could add commands that provide finer detail. For example, if each of the three patterns we were looking for consisted of six bytes, we might put a value in the table as each byte was received. Then we could see where in the detection of pattern 2 the unmask command was received. If the unmask command was received after the first or second byte of the pattern, we might suspect a bug in the code that prevents unmask commands from being properly handled if they are received during the pattern they attempt to mask (possibility 2 of the three listed above). On the other hand, if the unmask command is received just before the last byte of the pattern, we might suspect that the PC is too slow (possibility 1) or a race condition (possibility 3).

Of course, if we expand this too far, we fill our trace buffer with so much information that the data we need might be overwritten. One way around this is to add data to the buffer instead of adding only codes. For example, we might have a 16-bit free-running timer and we might grab the count in the timer and store it with each action code. That would timetag each value in the buffer and give us some idea of when it occurred. Let's say that our timer runs at 100 Khz, so each count is 10 microseconds. If our diagnostic code stores the 16-bit count each time it captures a value, the last few values in our trace buffer might look like this:

64 FF 03 64 12 8B 06 14 23 16 15 14 07 15 15 FF

Now, we have a problem! With two FF values in the table, how do we tell which one is the end? In this case, since we derived this example from the previous one, we know what the values are, and we can rewrite the data as follows:

Code 64 (Interrupt) at time FF03

Code 64 (Interrupt) at time 128B

Code 06 (Start of seq 2) at time 1423

Code 16 (Unmask seq 2) at time 1514

Code 07 (Seq 2 detected, but masked) at time 1515

Okay, we've deciphered the contents of the buffer, but what do the time values mean? Since the code is reading the contents of a free-running, 16-bit counter, the time values are a count. We said before that each count is 10 microseconds of resolution. So let's take the FF03 value as our zero seconds timebase. The 128B value obviously occurred after the counter rolled over. So the difference between 128B and FF03 is 10000 (hex) – FF03 + 128B, or 1388 (hex), or 5000 decimal. Since we have a 10-microsecond timebase, the two interrupts occurred 5000×10 µs, or 50 milliseconds apart. This is not surprising, since that is supposed to be the interrupt period.

The second value in our table is easier, since there was apparently no rollover, so the time is 10 µs × (1423 – 128B), or 4.08 ms. Doing the same thing with the rest of the values, we can rewrite our trace table as follows, including both the absolute time and the time relative to the previous tracepoint:

	Absolute	**Relative**
Code 64 (Interrupt)	0	0
Code 64 (Interrupt)	50 ms	50 ms
Code 06 (start of Seq 2)	54.08 ms	4.08 ms
Code 16 (Unmask Seq 2)	56.49 ms	2.41 ms
Code 07 (Seq 2 detected, masked)	56.5 ms	10 µs

Seeing this time scale, we would probably assume that our problem was a race condition or a slow PC, since the unmask code occurred one count before the "pattern detected" action code.

Let's look again at the end-of-table problem. One way around the problem would be to insert 3 bytes of FF at the end of the table (FF FF FF). Assuming you have no action codes that are FF, you can always find the end of the table by looking for this pattern. Since we are (in this example) storing an action code followed by a word of data, the FF FF FF pattern will never occur except at the end of the table.

The data associated with an action code does not have to be a time-tag. You could write the value of a pointer, a value captured from an A/D converter, or any other meaningful number that goes with the specific action code. The key to making this work is to be sure that all action codes write the same amount of data—a code and two additional bytes, in our example. Even if some codes don't need additional data, write zeros or something anyway. If you don't, it will be nearly impossible to determine where the codes are, in a design of any complexity. Quick example: in the trace we've been looking at, how do you tell whether the last value in the table (15) is the second byte of a timetag or a valid action code?

Another way around the problem of finding the end of the table is to use two tables. With this method, one table holds the action codes, and the other table holds the timetag or other word data. Obviously, the data table will be twice as large as the action code table. If each table uses the same pointer (multiplied by 2 for the data table), the pointers will always be in sync. Our hypothetical example would then look like this:

Code Table: 64 64 06 16 07 FF

Data Table: FF03 128B 1423 1514 1515 FFFF

In some cases this is easier to read, since the action codes are all grouped together, giving an easy-to-read sequence of events. However, in the cases where you need to look at what is in the data table, you have to switch back and forth, keeping track of what data value goes with which action code.

The obvious drawback to trace buffers is that they take time to execute and memory to store the information. Adding timetags or other data takes more time and more memory. There are other things to watch for when using a trace buffer.

Multiple Rollovers

When you are timetagging data using a free-running counter, remember that a rollover (such as we had with the first two data points in our example) may be more than one rollover. That is, the count may have rolled over, gone to FFFF, and rolled over again. In most debugging scenarios, you will care only that the difference is big, not how big it is. If you really need to know the exact time, there are some ways around this problem. One is to use a counter with more bits, enough that it will never roll over between two events. Another solution, if you are using a regular timer tick interrupt (such as the 50 ms tick in the example), is to reset the counter at every timer tick. The count is then the time from the last tick. A final method, if your counter can generate interrupts on rollover, is to make the timer rollover interrupt generate a unique action code. Then every rollover is noted in the table.

Changing Count

If you read the timetagging counter while it is changing, you may get a bogus value. This is more likely to happen if you are using a counter IC that has 16-bit (or wider) counters, but an 8-bit interface. If the count rolls over from, say, 40FF to 4100, you might read 41FF if you read the low byte first (or 4000 if you read the high byte first). Some counters will allow you to "freeze" the count in a latch so you can read it without worrying about this problem. Another solution is to read the count multiple times until you get two that agree.

Count Ambiguity

In the example we looked at, the final time value was one count larger than the previous data point (1514 versus 1515). We assume this means that the time was one count, or 10 microseconds. However, the actual time could have been anywhere from 0 to 20 microseconds. It works like this: if the time read for the next-to-last data point (action code 16) occurred just before the timer incremented to 1515, and the time read for the last data point (07) occurred just after that increment, the time could be zero (or as close as possible, given software latencies). On the other hand, if the time read for the 16 action code occurred just after the counter incremented to 1514, and the time read for the 07 action code occurred just before the counter incremented to 1516, then there would be nearly two full counts, or 20 microseconds, between the two events. This is not only a potential problem for timetagging situations, it can also be a problem any time a counter is used to time events. Figure 2.2 illustrates the ambiguity.

Buffer Overflow

In our hypothetical example, there was a timer interrupt that was relatively slow when compared with the data rates and other actions involved. If you have a faster timer, or other events that produce action codes that occur infrequently, you may find a trace buffer that contains nothing but timer tick action codes. In this kind of system, you might want to leave the timer out of the trace buffer, or provide a variable you can set during debug to enable and disable storing of that action code.

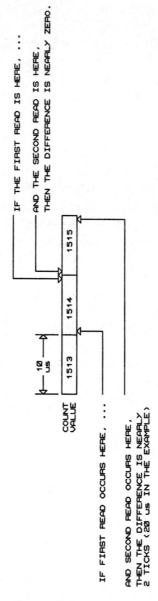

Figure 2.2 Timer Ambiguity

Hardware Trace Outputs

In some systems, such as a small microcontroller-based design, there isn't enough memory to implement a trace buffer, or there isn't a serial port to transfer data back and forth to a PC. In these cases, you can get much the same effect by outputting the data to an unused memory or I/O location and capturing it on a logic analyzer in state mode. The analyzer memory serves as the trace buffer. Figure 2.3 shows how this could be implemented in our hypothetical data stream monitor circuit, if it were based on an 80188 and used I/O decoding similar to that of Cody's programmer. The timing of the data versus the TRACE WR signal is also shown in the figure.

As you can see from the figure, the 74ACT138 decodes the low order address (latched with the 74ACT373), is enabled with PCS2 from the 80188, and with WR from the 80188. On the 80188, the PCSx lines are active for 128 contiguous addresses and may be mapped into either memory or I/O space. Let's say for now that PCS2 is active (low) for all I/O addresses in the range 100–17F (hex, 512–639 decimal). Y0, Y1, and Y2 are used; the remaining lines are free. Y7 has been connected to a signal called TRACE WR, which doesn't connect to any hardware. Because of the decoding, Y7 is active at addresses 138 to 13F (hex). If the software were to write action codes to this address, it would not affect any hardware, but a logic analyzer could pick up the codes and store them. Figure 2.4 shows how a logic analyzer might be connected to capture these codes, and how the display might look for the first 15 codes from the data stream monitor. Compare these with the trace buffer values we looked at earlier.

Of course, we might want additional information in some cases, just as we did with the trace buffer. One advantage to using a logic analyzer is that we don't have to keep track of memory pointers, so we can have some action codes that have data (such as a timetag) associated with them, and others that use just the action code alone. Most modern logic analyzers can timestamp captured states, eliminating the need for the software to do this, and eliminating the counter we used in a previous example for this purpose. Of course, if the data captured comprise anything other than time, we still need a way to tell the difference between action codes and data.

It is clear from the partial schematic shown in Figure 2.3 that the ACT138 decodes only A3, A4, and A5 (called a partial decode) of the address bus, so each line is active for more than one address. For example, Y0 (pin 15) is active for addresses 100 to 107, and again from 140 to

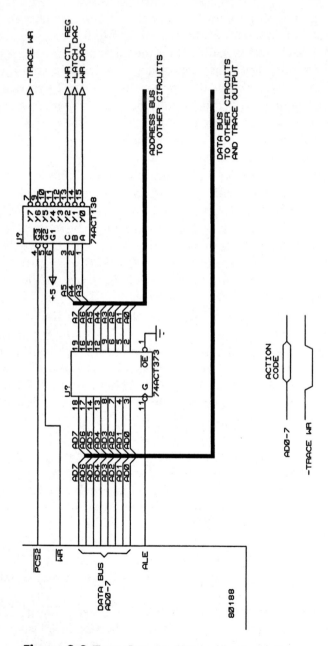

Figure 2.3 Trace Outputs Using a Logic Analyzer

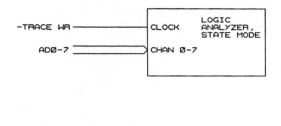

```
                                      LOGIC
 -TRACE WR ──────────── CLOCK     ANALYZER,
                                      STATE MODE
    AD0-7  ══════════╗  CHAN 0-7
```

```
   STATE   POD  1
           HEX

   0000     64
   0001     64
   0002     03
   0003     04
   0004     64
   0005     03
   0006     04
   0007     15
   0008     16
   0009     17
   0010     64
   0011     03
   0012     05
   0013     64
   0014     06
```

Figure 2.4 Connecting a Logic Analyzer to Capture Action Codes

147, 180 to 187, and 1C0 to 1C7. The following table shows the address range for which each decode is active. Remember that, on the 80188, PCSx lines are active for 128 contiguous addresses. Figure 2.5 shows how we would connect the logic analyzer to take advantage of this.

Address	Active ACT138 Output	Also Active at:
100–107	Y0 (pin 15)	140–147, 180–187, and 1C0–1C7
108–10F	Y1 (pin 14)	148–14F, 188–18F, and 1C8–1CF
110–117	Y2 (pin 13)	150–157, 190–197, and 1D0–1D7
118–11F	Y3 (pin 12)	158–15F, 198–19F, and 1D8–1DF
120–127	Y4 (pin 11)	160–167, 1A0–1A7, and 1E0–1E7
128–12F	Y5 (pin 10)	168–16F, 1A8–1AF, and 1E8–1EF
130–137	Y6 (pin 9)	170–177, 1B0–1B7, and 1F0–1F7
138–13F	Y7 (pin 7)	178–17F, 1B8–1BF, and 1F8–1FF

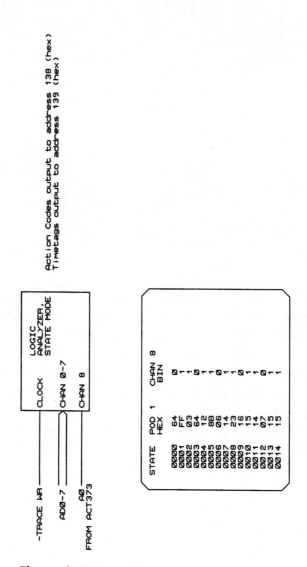

Figure 2.5 Expanding Action Codes Using a
Logic Analyzer

Again, we have used Y7 as our diagnostic strobe and have connected it to the clock on the logic analyzer. We still capture the data on pod 1 of the analyzer. But we have also added A0 from the 80188 address latch to the logic analyzer, connecting it to Channel 8. Now, when we want to write an action code, we output it to address 138 (hex) as we did before. But when we want to output additional diagnostic data, we output it to address 139 (hex). Figure 2.5 shows the resulting display, using as an example the timetag values we used before in the data stream monitor trace buffer.

Writing to address 138 toggles the TRACE WR signal, clocking data into the analyzer as before. But writing to address 139 also toggles the TRACE WR signal, also clocking the data into the analyzer, but with the A0 line high. Now we can tell the difference. We can even accompany some action codes with more than 2 bytes of data, because we can always find the action code corresponding to a stream of data bytes by looking backward in the analyzer buffer for the write with A0=0.

We could do the same thing by connecting A6 instead of A0 to the analyzer and writing the action codes/data to 138/178.

Other Methods of Generating Trace Data

This general concept will work with any processor that uses an external data bus and decodes the address to generate write strobes to external hardware. Sometimes there is no address decode logic, or all the strobes are used, or some other situation prevents you from using a decoded write strobe to trigger the analyzer. In this case, there are other tricks you can use to generate a strobe for the trace data.

Write to ROM

In most embedded systems, the only thing in the ROM space is the ROM itself. However, the address decoding logic often does not differentiate between a read and a write. If you write your trace data to the ROM space, it usually will not affect any hardware. Use caution, though: make sure the address decode logic doesn't turn on the ROM outputs when you write to that address range, or you will have bus contention. If you are using a nonvolatile memory such as an EEPROM, be sure a write to the ROM space will not actually change the contents of the memory. Of course, if you have debug RAM in the ROM space, this technique is unsuitable.

Figure 2.6 shows the processor, PROM, and decoding logic for an 80188 system such as Cody's programmer. The logic gates circled on the (partial) schematic decode a write to the ROM space, and would generate a strobe to clock trace data into an analyzer. Like the decoded write strobe, the trace clock is labeled TRACE WR. The actual circuitry could be a permanent part of the board, or could be connected to the circuit with a DIP clip when needed. When using the write-to-ROM scheme for Motorola-type processors that need an ACK signal to terminate the cycle, make sure the PROM decoding logic generates the ACK on both reads and writes.

Read from ROM

On some microcontrollers, such as the 8031, you cannot write to the ROM space. In these cases, if you are not using all the ROM space, you can do the same thing by reading from the ROM. In this case, the trace data is taken from the low order *address* lines. Figure 2.7 shows an 8031 with logic to detect and generate a write strobe on read-from-ROM. In this case, the circuit uses a 27256 EPROM, which requires 32k of the 64k address space. We decode any read from the upper 32k of the address space (A15=1) as a trace output.

The code to implement this read-from-ROM technique is shown below.

Test Header

I like to put a 10-pin inline or dual-row header on boards for the trace output signals. Eight pins are for the data lines (16, if I'm using a 16-bit processor), one for the write strobe, and one for ground. On some designs where cost has to be minimized, you could do the same thing by adding an edge connector on one side of the board. In one case, where board real estate was at a premium, I just added a row of pads and accessed the test points with a fixture containing spring-loaded "pogo pins" of the type used in bed-of-nails testers.

Listing 2.2 Read-from-ROM for trace output on 8031

```
                     ; Code fragment to implement trace
                     ; write using Read-from-ROM on 8031.
MOV DPTR,#8000h      ; DPTR points to trace write addr
MOV A,#TRACEVALUE    ; reg A contains trace value to output.
Mov A,@A+DPTR        ; Output the trace data.
```

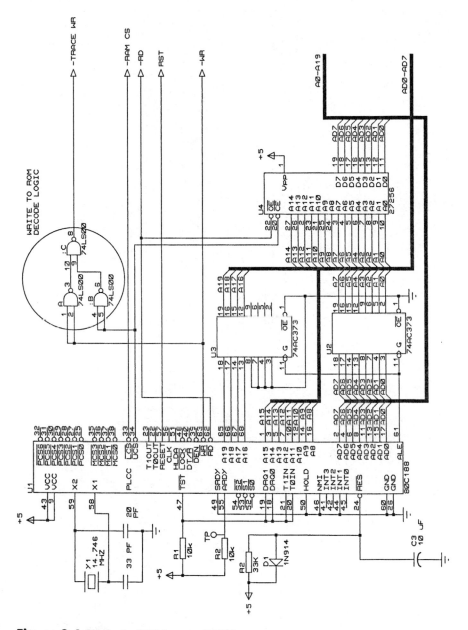

Figure 2.6 Write-to-ROM on an 80188

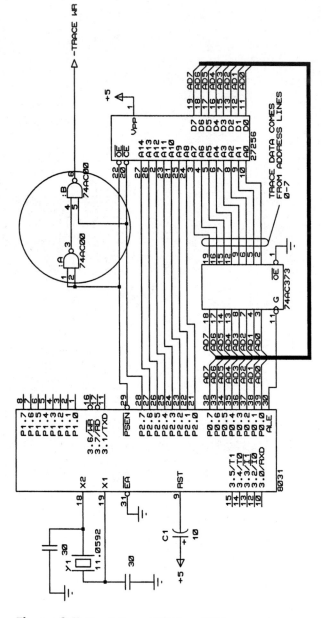

Figure 2.7 Read-from-ROM on 8031

LEDs

It seems that nearly every embedded designer uses LEDs as status indicators. These are typically driven from a register output or from a port pin, as shown in Figure 2.8.

Three LEDs are shown in Figure 2.8. Connected to pin 8 of the 8031 microprocessor through a MOSFET is a single LED. The drive transistor is needed on some processors because the sink current available from the processor output is insufficient to provide adequate brightness of the LED. A similar LED drive is connected to pin 19 of the 74AC374. If a bipolar LED is used, as shown connected to pins 2 and 5 of the 74AC374, you get some interesting possibilities, as shown by the table below:

Bipolar LED Drive 74AC374 Pin 5	74AC374 Pin 2	LED Color
0	0	Off
0	1	Green
1	0	Red
1	1	Off

By toggling quickly between the 10 and 01 states, an amber color can be obtained.

When a system has a single software-controlled LED, it is often used as a good/bad indicator. This can be useful to indicate which board to replace, but you often want a more detailed idea of what went wrong. One way to do this is to flash an error code on the LED. Listing 2.3 shows how this can be done on an 8031. This code turns the LED on for about one second as a start indicator, then flashes the LED quickly to indicate the most significant digit and more slowly to indicate the least significant digit. To read the error code, you would look for the start indication, count the slow flashes, and then count the fast flashes for a two-digit error code. Since the user has to remember the error count, I'd recommend that neither digit have a value greater than 5. Of course, the technique shown here will work with any processor. This 8031 code would be suitable as a display-and-hang error handler. If you want the processor to keep running, servicing interrupts, or talking to a display, you could implement the LED delays as part of a regular tick interrupt service routine.

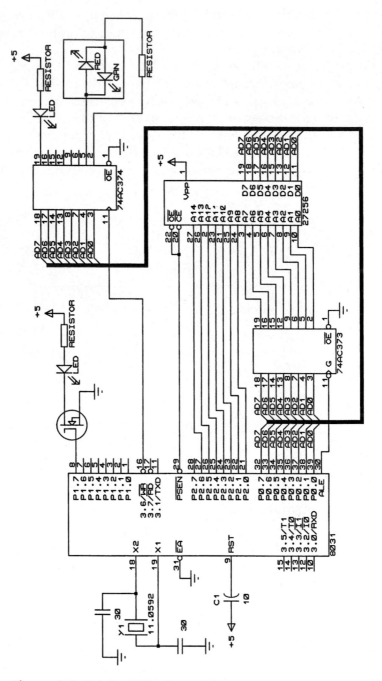

Figure 2.8 Driving LEDs from a Microprocessor

Listing 2.3 Flashes a Two-Digit Error Code on an 8031

```
; 8031 error-code flasher.
; Code to flash is in R0.
; Flashes one long (1 sec) ON time,
; then MSB at about .4 sec/digit, then
; LSB at about .8 sec/digit.
; So a code of 53 results in a
; 1 sec ON time, then five quick flashes
; then three slow flashes.

; A '1' at P1.0 turns the LED ON, a '0'
; turns it off.
; The input crystal is 11.0592 MHz, and we
; use timer T0 to generate the flash rate.
; R1 and R2 are used as temp storage locations.
flash:
; First, break the two-digit code down into
; a msb count (in R2) and an LSB count (in R1).
     mov a,r0
     anl a,#0fh
     mov r1,a
     mov a,r0
     rr a           ; get msb to lsb.
     rr a
     rr a
     rr a
     anl a,#0fh   ; strip off high bits.
     mov r2,a
     ; Now r2 — upper 4 bits of orig value,
     ; r1 contains lower 4 bits.

     ; To output a 1 sec ON time, we need 14 iterations
     ; of 65536 counts.

     clr p1.0
     mov a,#14
startoff:
     call delay
     dec a
     jnz startoff

     setb p1.0    ; turn on LED.
     mov a,#14    ; 14 cycles.
startset:
     call delay   ; delay one count of FFFF.
     dec a
     jnz startset ; loop until a=0.
```

Listing 2.3 Flashes a Two-Digit Error Code on an 8031

```
      ; Now send the MS count. This is done by
      ; turning the LED on and off for six intervals,
      ; decrementing R0, and repeating until R0
      ; goes to 0.

      ; Since the start bit was on, do off first.
msoff:
      clr p1.0
      mov a,#3    ; flash on time = 3.
msofflp:
      call delay
      dec a
      jnz msofflp
      ; Off time done, turn on LED and do ON time.
      setb p1.0
      mov a,#3    ; flash off time = 3.
msonlp: call delay
      dec a
      jnz msonlp ; loop until done.

      ; ON time done, now decr MS bit count,
      ; repeat until done.
      mov a,r2
      dec a
      mov r2,a
      jnz msoff  ; loop until r0 = 0.

      ; Now turn the LED off and delay
      ; to make a break between the
      ; fast and slow flashes.
      clr p1.0   ; LED off
      mov a,#3
      call delay

      ; Ms nybble done, now output ls nybble at
      ; slow rate (12 delays, 6 on, 6 off).

lsoff:
      clr p1.0
      mov a,#6    ; on time = 6.
lsofflp:
      call delay
      dec a
      jnz lsofflp
      ; Off time done, turn on LED and do ON time.
      setb p1.0
      mov a,#6    ; off time = 6.
```

Listing 2.3 Flashes a Two-Digit Error Code on an 8031

```
lsonlp:
    call delay
    dec a
    jnz lsonlp ; loop until done.

    ; ON time done, now decr LS bit count,
    ; repeat until done.
    mov a,r1
    dec a
    mov r1,a
    jnz lsoff  ; loop until r1 = 0.

    jmp flash  ; keep outputting the code forever.

    ; Delay is a support routine that delays for a
    ; count of 65536 (71 ms). Uses T0, no regs.
delay:
    clr tf0
    mov th0,#0
    mov tl0,#0
    mov tmod,#1
    setb tr0
delaywait:
    jnb tf0, delaywait ; loop until timer done.
    clr tr0
    ret
```

Serial Condition Monitor

If you had a microcontroller design that had only one pin free for diagnostics, you might decide to implement a serial condition monitor. In this scheme, you output a start bit (needed to always trigger the scope in the right place), then output the data 1 bit at a time. You determine what the status of the software is by which bits are turned on. Figure 2.9 illustrates this.

The serial condition monitor typically sends bit-oriented information rather than a trace output byte. This is because, first, if you display 8 bits on a scope screen, it is difficult to decode which ones are set. Also, the serial condition monitor cannot be generated just any time, or the 'scope will have difficulty triggering. The serial condition data would typically be output once each time the code goes around the background loop, or every time a regular timer interrupt is serviced.

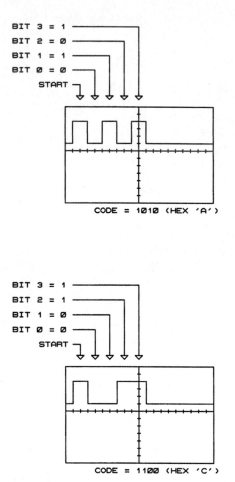

Figure 2.9 Serial Condition
Monitor

Using the data monitor system from earlier in this chapter as an example, we might define the bits as follows:

Bit 0: 1 = Pattern 1 masked, 0 = unmasked

Bit 1: 1 = Pattern 2 masked, 0 = unmasked

Bit 2: 1 = Pattern 2 masked, 0 = unmasked

Bit 3: 1 = Data stream active, 0 = no data being transferred

Other examples might include a bit to indicate when there is data in a receive or transmit FIFO, when a motor is active, when one system is waiting for another, and so on. The serial condition monitor technique is not very well suited to displaying events that come and go very quickly—they are too hard to see on a 'scope.

Typically, the code that outputs the serial information will send part of a byte; other routines set and reset bits in the byte to indicate status. The two code fragments below show how this technique can be implemented on an 8031 and on a PIC17C42.

Listing 2.4 PIC17C42 Assembly Code for Serial Status Output

```
; STATUS OUTPUT TO PORT D BIT 0.
; PORT D BIT 0 MUST BE CONFIGURED AS OUTPUT.
; CODE SENDS LEAST-SIGNIFICANT FOUR BITS OF
; BYTE DIAG TO PORT.
; OUTPUT SEQUENCE IS:
; SYNC BIT (1)
; DIAG BIT 0
; DIAG BIT 1
; DIAG BIT 2
; DIAG BIT 3
; ZERO
; EACH BIT TAKES ABOUT 1.6 MICROSECONDS ON A PIC17C43
; WITH A 12-MHZ CRYSTAL.

        MOVLB 1
        BSF PORTD,0  ; OUTPUT SYNC BIT (1).
        NOP
        NOP
        BTFSC DIAG,0 ; DIAG BIT 0 SET?
        GOTO MSET0   ; YES, OUTPUT A 1.
        BCF PORTD,0  ; NO, OUTPUT A 0.
        GOTO M1
MSET0:
        BSF PORTD,0
        NOP
M1:
        BTFSC DIAG,1 ; DIAG BIT 1 SET?
        GOTO MSET1   ; YES, OUTPUT A 1.
        BCF PORTD,0  ; NO, OUTPUT A 0.
        GOTO M2
MSET1:
        BSF PORTD,0
        NOP
M2:
        BTFSC DIAG,2 ; DIAG BIT 2 SET?
        GOTO MSET2   ; YES, OUTPUT A 1.
```

Listing 2.4 PIC17C42 Assembly Code for Serial Status Output *(continued)*

```
      BCF PORTD,0  ; NO, OUTPUT A 0.
      GOTO M3
MSET2:
      BSF PORTD,0
      NOP
M3:
      BTFSC DIAG,3 ; DIAG BIT 3 SET?
      GOTO MSET3 ; YES, OUTPUT A 1.
      BCF PORTD,0 ; NO, OUTPUT A 0.
      GOTO MSDON
MSET3:
      BSF PORTD,0
MSDON:
      NOP
      NOP
      BCF PORTD,0 ; TERMINATE WITH A 0.
      MOVLB 0 ; END OF DIAGNOSTIC OUTPUT.
```

Listing 2.5 8051/8052 Assembly Code for Serial Status Output

```
; THIS CODE FRAGMENT OUTPUTS A FOUR-BIT STATUS
; VALUE, DIAGNOSTIC, TO PORT 0 BIT 0 OF AN
; 8051/8052 PROCESSOR. OUTPUT SEQUENCE IS:
; START BIT (1)
; DIAGNOSTIC BIT 0
; DIAGNOSTIC BIT 1
; DIAGNOSTIC BIT 2
; DIAGNOSTIC BIT 3
; ZERO.
; ON AN 8051 WITH AN 8-MHZ CRYSTAL,
; EACH BIT WILL BE ABOUT 4.4 MICROSECONDS LONG.

      MOV ACC,DIAGNOSTIC
      SETB P0.0      ; OUTPUT START BIT (1)
      RRC A
      MOV P0.0,C     ; OUTPUT DIAG BIT 0
      RRC A
      MOV P0.0,C     ; OUTPUT DIAG BIT 1
      RRC A
      MOV P0.0,C     ; OUTPUT DIAG BIT 2
      RRC A
      MOV P0.0,C     ; OUTPUT DIAG BIT 3
      NOP
      NOP
      CLR P0.0       ; ALL DONE.
```

Bit-Banging Serial Interface

Software monitor programs need a serial port to operate. The problem is, in many designs, there is only one serial port available, and you need that one for the design itself. The 8031 is a good example of a design that has only one serial port available. If your design uses a microprocessor that does not have an on-chip serial port, you may not want to add a hardware UART just for debugging.

It is possible to implement a serial port in software. Each byte of RS-232 serial data is transmitted as a low-level start bit, 8 data bits (LSB first), then a high-level stop bit (assuming no parity). These levels are at the logic level; they are inverted before going to the RS-232 connector. If you are using a microcontroller and have a spare port pin, or if you have a spare register output bit on a multichip processor design, you can implement a serial port in software. Listing 2.6 shows a routine

Listing 2.6 Post-Mortem Memory Dump on 8031 Using Software UART

```
; 8031 post-mortem dump, sending data at
; 9600 baud to port P1.0 using bit-banging.
; The input crystal is 11.0592 MHz, and we
; use timer T0 to generate the baud clock.
;
; All locations from 0 to 7F are sent.
; Output format is two hex digits, one space,
; cr/lf after every 16 bytes.

init:                   ; Initialize timer 0.
      mov th0,#159 ; 255-159 = 104 us = 9600 baud.
      mov tl0,#0
      mov tmod,#2   ; set timer 0 for mode 2
                    ; (8-bit reloading, free running).
      setb tr0      ; enable timer0.

; Read contents of memory from 00-7F, convert to ASCII,
; and output via bit-banging serial routine.
; Uses R0, acc, DPTR.
dumploop:
      mov a,@r0
      rr a
      rr a
      rr a
      rr a
      anl a,#0fh
```

Listing 2.6 Post-Mortem Memory Dump on 8031 Using Software
UART *(continued)*

```
    mov dptr,#asctable ; convert hi nybble to ascii.
    movc a,@a+dptr
    call aout          ; xmit lo nybble
    mov a,@r0
    anl a,#0fh
    mov dptr,#asctable ; convert lo nybble to ascii
    movc a,@a+dptr
    call aout          ; xmit lo nybble of byte
    mov a,#' '         ; space after digit pair
    call aout
    inc r0             ; Incr memory pointer
    mov a,r0
    xrl a,#080h        ; all done?
    jz done            ; yes, exit.
    mov a,r0           ; no, check for EOL
    anl a,#0fh         ; EOL (every 16 bytes)?
    jnz dumploop       ; no, keep looping.
    mov a,#0dh         ; yes, send cr, lf every 16 bytes
    call aout
    mov a,#0ah
    call aout
    jmp dumploop
done: jmp done
asctable:          ; hex-to-ascii table
    db '0123456789ABCDEF'

aout:      ; Sends contents of accum to
           ; port 1.0 by bit-banging.
           ; Uses no other registers.
startbit:
    jnb tf0,startbit ; loop until overflow
    clr tf0
    clr p1.0
bit0:
    jnb tf0,bit0
    clr tf0
    rrc a
    mov p1.0,c
bit1:
    jnb tf0,bit1
    clr tf0
    rrc a
    mov p1.0,c
bit2:
    jnb tf0,bit2
```

Listing 2.6 Post-Mortem Memory Dump on 8031 Using Software UART *(continued)*

```
        clr  tf0
        rrc  a
        mov  p1.0,c
bit3:
        jnb  tf0,bit3
        clr  tf0
        rrc  a
        mov  p1.0,c
bit4:
        jnb  tf0,bit4
        clr  tf0
        rrc  a
        mov  p1.0,c
bit5:
        jnb  tf0,bit5
        clr  tf0
        rrc  a
        mov  p1.0,c
bit6:
        jnb  tf0,bit6
        clr  tf0
        rrc  a
        mov  p1.0,c
bit7:
        jnb  tf0,bit7
        clr  tf0
        rrc  a
        mov  p1.0,c
stopbit:
        jnb  tf0,stopbit
        clr  tf0
        nop
        setb p1.0
        ret
```

that will transmit the first 128 bytes of memory to a PC using this technique, which is sometimes called *bit-banging*. The code in the listing converts each byte in memory to two hex digits and separates each pair of digits with a space. A carriage return/linefeed pair is added after every 16 digits.

This routine, or similar code on another processor, could be used to implement a post-mortem dump of the memory contents after a failure.

It might be activated by a hardware reset (manually initiated when an error occurs), or an unused interrupt could be used to trigger the dump.

The code as shown uses one of the 8031 internal timers. You could implement the code using software loops for timing. This would be fairly straightforward on a processor such as the 8031; on more complex processors, with independent bus interface and execution cores, the timing is less predictable and a hardware timer is easier to use.

It is also possible to implement a serial receive routine in software, although that is probably less useful. You can construct a complete monitor program around these routines, although they have some limitations. The processor cannot do anything else while receiving or transmitting, and the interface is half duplex—you cannot receive and transmit at the same time.

On a very fast processor, it is possible to implement a full-duplex UART in software by having a regular timer interrupt that operates at a 4x, 8x, or other multiple of the baud rate. I did this once with a DSP for a special test fixture, but the fast interrupts really tie the processor up.

Listing 2.7 shows a software serial receive routine implemented on an 8031. The routine is intended to function as an ISR, with the serial data connected to the 8031 INT0 interrupt pin. INT0 would be programmed to be edge-sensitive. When the leading edge of the start bit occurs, the ISR reads the serial data. Note that if there are other, higher priority, interrupts used in the system, the interrupt latency for the serial receive code may be too long for reliable operation.

Listing 2.7 Serial Receive Routine for 8031

```
; Swuartin is the receive process. It is activated
; by an interrupt on INT0, (P3.2) which is the
; leading edge of the start bit for the incoming
; character.
; The code sets up timer 0 as an 8-bit reloading
; timer, sampling approximately in the middle of the
; bit. This code expects rx data to be 8N1.
; Code assumes a crystal frequency of 11.0592 MHz.
; Not shown is a routine, ERROR, which must handle
; bad start bits and any other errors that may occur
; in the serial data stream. Could be left out
; for a debugger application, if the start bit
; isn't verified at the first timer rollover.
swuartin: ; Initialize timer 0.
```

Listing 2.7 Serial Receive Routine for 8031 *(continued)*

```
        clr tr0         ; Stop Timer0 in case it's running
        mov th0,#160 : 255-160 = 104 us = 9600 baud.
        mov tl0,#210 ; Starting at 210 will put the
                        ; first sample about in the
                        ; middle of the start bit.
                        ; Subsequent samples will be
                        ; spaced 1 bit time apart.
        mov tmod,#2  ; Set timer 0 for mode 2
                        ; (8-bit reloading, free running).
        clr tf0         ; Clear any pending rollover indication
        setb tr0        ; Enable timer0.

rxstart:
        jnb tf0,rxstart  ; Wait for middle of bit.
        jb p3.2,error    ; If P3.2 is set, then we had
                         ; an invalid start bit.
        clr tf0
        mov a,#0         ; Clear byte accumulator.
rxd0:
        jnb tf0,rxd0     ; Wait for middle of bit.
        clr tf0
        mov c,p3.2       ; Get the incoming bit,
        rrc a            ; rotate into the byte accumulator.
rxd1:
        jnb tf0,rxd1     ; Wait for middle of bit.
        clr tf0
        mov c,p3.2       ; Get the incoming bit,
        rrc a            ; rotate into the byte accumulator.
rxd2:
        jnb tf0,rxd2     ; Wait for middle of bit.
        clr tf0
        mov c,p3.2       ; Get the incoming bit,
        rrc a            ; rotate into the byte accumulator.
rxd3:
        jnb tf0,rxd3     ; Wait for middle of bit.
        clr tf0
        mov c,p3.2       ; Get the incoming bit,
        rrc a            ; rotate into the byte accumulator.
rxd4:
        jnb tf0,rxd4     ; Wait for middle of bit.
        clr tf0
        mov c,p3.2       ; Get the incoming bit,
        rrc a            ; rotate into the byte accumulator.
rxd5:
        jnb tf0,rxd5     ; Wait for middle of bit.
        clr tf0
```

Listing 2.7 Serial Receive Routine for 8031 *(continued)*

```
      mov c,p3.2          ; Get the incoming bit,
      rrc a               ; rotate into the byte accumulator.
rxd6:
      jnb tf0,rxd6        ; Wait for middle of bit.
      clr tf0
      mov c,p3.2          ; Get the incoming bit,
      rrc a               ; rotate into the byte accumulator.
rxd7:
      jnb tf0,rxd7        ; Wait for middle of bit.
      clr tf0
      mov c,p3.2          ; Get the incoming bit,
      rrc a               ; rotate into the byte accumulator.
rxstop:
      jnb tf0,rxstop      ; Wait for middle of bit.
      jnb p3.2, error     ; If P3.2 not set, invalid stop bit.
      clr tf0
; At this point, the received character is in
; the accumulator.
```

Debug-Only Serial Interface

Sometimes you want a full serial interface for debugging, but you cannot justify adding the hardware to every production board just in case it's needed. An alternative approach is to connect the processor data bus and control lines to a pin header. You can then build a single serial interface daughterboard containing a UART and RS-232 interface. The serial interface board connects to the pin header, and you plug it in only when you need to debug a problem. If you have a spare interrupt available for the serial interface, you can leave a ROM-based debugger resident in your program (but be careful of license fees!) and activate it when a character is received via the serial interface. If you pull the interrupt to the inactive state on the processor board, the software should never see an interrupt unless the serial board is installed. Figure 2.10 shows this concept in block diagram form. Another means of testing for the serial board is to have a bit, testable by the CPU, that is pulled up on the processor board, connected to the header connector, and grounded on the serial board. However you do it, you don't want the software to execute the debugger code unless the serial board is installed.

In the next chapter, we look at some general debugging tips.

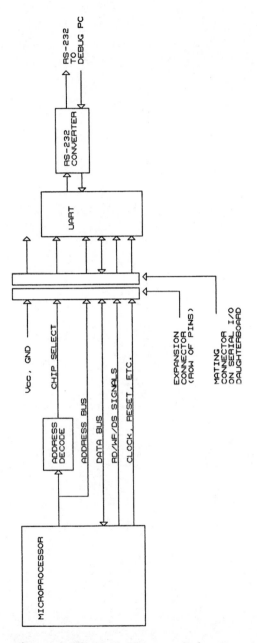

Figure 2.10 Block Diagram, Debug-only Serial Interface

3

General Debugging Tips

Cody waved three stapled sheets of paper at Matt. "Have you seen this ridiculous system test document that we've got to meet?" he asked.

"Seen it? I wrote most of it," Matt replied mildly.

"Oh." Cody replied. "Sorry."

"What's ridiculous about it?"

"I guess I just don't see the point. What's it for?"

Matt leaned his chair back. "Have you heard about the ClearScan project?"

"Not much. Just that it was some kind of disaster."

"We got that project from our sister division out in the beltway. Management there had convinced corporate that it was ready to go. The real story was that they just stopped work when the deadline arrived, but we didn't find that out until later. Corporate said we had to ship it because marketing had already sold it and there were penalty clauses if we didn't deliver. It fell apart in the field. The firmware locked up, lost data, and generally wouldn't do what it was supposed to do. Units got sent back, contracts got canceled, it was a real scramble. When we asked for the engineering and product verification test records, there weren't any. The design team had run a few cursory tests after the software seemed to work, then went on to other projects. By the time we had it in production, of course, they were all on some other project, so they didn't have time to fix it. Corporate finally decided to pull the product off the market. It got really messy, though."

"How so?"

"Well, to give you an example, I was putting together a presentation for one of our major customers who was looking at the ClearScan. We already had some indications of how bad it really was. Josh and I were reviewing the presentation with one of the marketing managers, and the guy from marketing said the customer wouldn't buy it with the MTBF numbers we had so far. He wanted us to make up some better numbers, and I said I wouldn't lie for them. Josh said that I just didn't understand the political realities of the situation."

"*Josh* said that?"

"He was under a lot of pressure. The company still thought the product could be salvaged, but it was up to Engineering to make it work."

Cody waved the papers again. "What's that got to do with this?" he asked.

"We wanted to be sure there wasn't a repeat of ClearScan, at least not from this group. We want procedures in place to ensure that our firmware-based systems are tested, that the tests make sense, and that the process is documented so we can trace it back if there are any questions. Larger projects have a formal Design Verification Procedure, but smaller projects like yours often slip through the cracks."

"Since you wrote it, can you answer some questions about it?"

"Sure."

Cody flipped to the second page and pointed to it. "I can figure out all the boilerplate stuff at the front, but what is this functional test plan, plan design review, and exception condition report?"

"You've got a specification for what your programmer is supposed to do, right?" Matt asked.

"Sure. You've seen it."

"How do you plan to be sure it does everything right?"

"Uh, try it with some of the target boards, I guess."

"That's exactly what we want to avoid. What you should do is make a list of all the functions you have to implement. Then you should write a plan that describes how you are going to verify those functions. You think of all the things that can go wrong in the software, like counters that roll over from zero to all ones, real-time clocks that roll over at midnight, operators hitting unexpected key combinations, everything. Then you add tests for those conditions to the plan, or an explanation of why the test is unnecessary. Then we have a design review, where you give your code and your test plan to other people, and they see if they can find anything you missed. Then we have a meeting, decide which ones are valid, and add them to the test plan. Then you go test it."

"Do we really do this all the time?"

"No. Very few processes are followed to the letter all the time, and sometimes there are valid exceptions when this process is unnecessary. This is just a goal, intended to make our designs better."

"Sounds like a lot of work."

"Fixing a software bug at some site in the Arizona desert is a lot more work."

Debugging: General Strategies

While Cody writes his test plan, we'll look at some general debugging strategies for embedded systems. There are differing philosophies about debugging, and there is no single correct way to do it. There are things you can do to simplify the process and make it less painful.

Embedded systems are a marriage of microprocessor hardware and software. Depending on the complexity of the design, the hardware may be a fairly simple "cookbook" design, or a nightmare of state machines and high-speed logic where the firmware plays a minor role in the real-time portion. In either case, the hardware must be working at a minimal level before software debugging can really begin. That is, if the processor cannot even get instructions from memory or write to I/O devices, there isn't really much the software can do.

The term *debugging* is a bit unfortunate at the system level of a design. A better phrase would be *design verification*. The debugging process should really be one of proving that the design does what it is supposed to do. There are two methods of doing this: the first one is to plug unproven code into hardware (proven or not), start it up, and see what happens. Bugs are tracked down and stomped one at a time. This method works, but it is prone to leaving undetected errors in the software.

A second method of debugging is to verify the hardware as far as is possible with simple routines. Then get a software framework going that can exercise the hardware. Finally, add and verify firmware functions one at a time. The only drawback to this method is that it represents a somewhat idealized picture of the real world. In most designs, you reach a point where you cannot add just one more function and test it. You have to turn on the interrupts, exercise real hardware in real time, and generally fall into the "wait-for-it-to-crash-then-examine-the-wreckage" mode of operation.

Once the software is working to the point that it runs without crashing, you have to verify the design. Of course, the first thing is to check that all the normal functions work. Be sure the display shows the right data at the right time and in the right place. Be sure all the keypad inputs work. Verify that correct input data produces correct motor motion. Verify that all this works, and works repeatedly.

After normal operation is verified, you need to check exception conditions. What happens if the user presses two keys at once? If he presses START right after pressing STOP? If the host computer sends garbage data over the RS-232 interface? You won't always think of all the ways an ingenious fool can mess things up, but you can at least find the obvious ones.

This process sounds very straightforward, but like most things in the real world, it is not as linear as it sounds. For example, if you are testing code modules, you might find it easier to verify the exception conditions for the serial input code while testing that module than waiting until the rest of the software is verified. Be careful, however, that whatever you change in fixing other bugs does not affect the performance of what you have already verified.

Although it seems like a lot of work, it is usually helpful to keep a logbook of the debug/verification process. This lets you record what you changed and why, what symptoms you fixed with each change, and even what hardware changes were made. When something that used to work quits working, you have a history of what was changed from then until now. Do not depend on your memory for this—you will forget. Even simple projects can benefit from a log of this type.

It's time for a story with a moral. There was a medical device, the Therac-25, that caused a number of patient deaths several years ago. The story was covered thoroughly in the *IEEE Computer* magazine and by other television and print publications, so the details will not be repeated here. The short version of the story is that the software was not able to cope with speedy operators. When certain key sequences were entered very quickly, the display showed that everything was all right, but the machine would deliver a lethal dose of radiation to the patient (there was also a second bug, of a different nature, that could cause similar results). In addition to the Therac-25, there has been at least one airliner crash has been linked to a combination of bad weather and an overlooked input condition in the fly-by-wire control system. Check the exception conditions, especially in any part of the code that has to do with safety.

Debugging Step by Step

Whether you use a functional verification approach or the plug-and-pray method, at some point you will encounter something that doesn't work—a bug. When this happens, you will know two things:

1. What you expected to happen.
2. What actually happened.

In some cases, the cause of the bug will be obvious. In others, it may not be clear what the software or hardware did wrong, or even which is at fault. In these cases, you should follow a procedure to isolate the fault. This procedure looks like this:

Form a hypothesis from the symptoms you observed.

Perform an experiment that will verify or disprove your hypothesis.

Evaluate the results and repeat the process as necessary to isolate the problem.

In some cases, you do not have enough information to form a hypothesis. In this case, your first experiment (or experiments) will have to be data collection exercises to gain additional information.

In debugging embedded systems, the "experiment" usually takes the form of setting a breakpoint *there* and looking at the registers and memory contents when it occurs, or of adding trace outputs, or something similar.

This procedure has been documented by other people, usually with the additional admonition that you should determine before running an experiment what you will do with the results. If your next step will be the same regardless of the result, then you can skip the experiment. This is a valid point, but it is based on an assumption that is often incorrect namely, that you know all the possible outcomes of the experiment. I have often seen a totally unexpected result from some kind of test, sometimes leading to a correct assessment of what is wrong.

Symptoms

Be sure you really know all the symptoms. Does the problem happen every time you run that particular sequence of inputs? If not, is the error regular or random? Is it really the START button that doesn't work, or is it all the keys in that row of the keypad?

One of the most common and most frustrating errors that I have seen in debugging is for someone to hang onto a theory about what causes a problem, even when the theory is contradicted by one or more symptoms. For example, consider the data monitor system from the previous chapter, which wasn't sending the pattern-detected indication even though it had supposedly been unmasked by the host. Suppose the hardware engineer suggests that the host PC is sending invalid codes to the data monitor system, so the data monitor doesn't see the unmask pattern. The software engineer displays the trace buffer. To refresh your memory, the last few bytes in the trace buffer looked like this:

Value	Meaning
64	20 ms interrupt service.
06	Pattern 2 start detected.
16	Unmask pattern 2 command received.
07	Pattern 2 verified, but masked by host.

A quick look at our trace data clearly shows that the data monitor system is seeing the unmask code, or at least some command that it is interpreting as an unmask code. At this point, all they can tell is that the unmask command was apparently received, but the data monitor still indicated that the pattern was masked when detected.

As we did in the last chapter, the software engineer adds timetags to the trace data, and now the trace buffer (still from the last chapter) looks like this (times are relative):

Value	Meaning	Timetag
64	20 ms interrupt service.	0
06	Pattern 2 start detected.	4.08 ms
16	Unmask pattern 2 command received.	2.41 ms
07	Pattern 2 verified, but masked by host.	10 μs

The engineers might theorize that the presence of the unmask code just prior to the detected-but-masked code is not a coincidence, and that the problem is caused by a race condition. If the same pattern shows up every time the error occurs, then they would be fairly safe in assuming that there actually is a race condition, caused either by a slow PC or by a bug in the unmask processing code (there are other possibilities, depending on the code structure, but we'll limit ourselves to what we have already discussed to avoid confusing the issue). On the other hand, if the unmask command sometimes occurs several milliseconds prior to the erroneous detected-but-masked code, then this would indicate that the problem is not a race condition (or that there is more than one cause for the symptoms).

As you can see, it is important not to wear blinders—if the evidence contradicts your theory, and even if the theory was your idea, you have to abandon the theory. Say that you have a design with a problem where processing seems to just stop for a couple of seconds. You theorize that some section of code is disabling interrupts and not reenabling them. But you notice that the heartbeat LED, driven by a timer interrupt, is still blinking. It is amazing how often people will hang on to a hypothesis, trying to figure out some other way the LED could keep blinking, or simply forgetting about it in their evaluation of the problem, instead of going with the facts. If you insist on holding on to your pet theory, explain what is wrong with the evidence. To stay with the LED example, maybe interrupts are not all disabled, but one critical interrupt is masked. Trying to explain away evidence is not always as silly as it sounds—sometimes in doing so you will think of something that didn't occur to you before. If you get results that seem impossible, based on your assumptions, *check your assumptions.*

Don't Shotgun!

Don't just randomly replace things, or change PLD equations, or modify the code, hoping the problem will go away. It probably won't. Worse, it might—just to reappear later. Understand the problem before fixing it. About the only scenario where shotgunning is useful is if there is a shorted IC and you cannot tell which one it is.

Test Equipment

Make sure your problem is not caused by your development and test equipment. I worked with an engineer on one problem that nearly drove us up the wall. It appeared that the processor simply was not executing certain instructions. The trouble turned out to be a conversion program in the development tool chain that had problems with a file format and dropped some data. Had we started with the assumption that our tools could be at fault, we would have saved some time.

Consider the Environment

Sometimes you have to suspect your environment. I worked on a system once that was distributed nationwide. Certain problems started popping up. After a while, we noticed that there was a pattern—they were worst in the winter, especially in northern states such as Michigan. What happens in the winter in Michigan? It gets cold. And dry. Bingo: static electricity. The machine had components, designed to be conductive just for this purpose, that were made wrong.

Now we rejoin Cody and Matt in debugging their programmer hardware.

4

Hardware Debug[*]

Matt saw Cody hunched over a bench as he strolled through the lab. As he approached, he could see a printed circuit board, populated with components, and an oscilloscope on the bench beside it. "How's it going?" he asked.

"It's not," Cody said. "It looks dead. The processor doesn't even run."

"What code have you got in the EPROM?"

"What do you mean?"

"Do you have special test code, or your application code?"

"Oh. The application code."

Matt frowned. "How are you going to tell if your problems are hardware or software?" he asked. "You've got unknown software running, or apparently, not running, on untested hardware."

"What would you suggest?"

"How about testing the hardware in small chunks, then working on your software when you've got some level of confidence in the hardware?"

Cody reached over and switched off the power supply that was connected to his board. "Okay," he said. "A divide-and-conquer strategy. I'm listening."

Matt picked up a pad of paper from the bench and handed it to Cody. "List the functional blocks from your block diagram," he said.

---

*Note: This chapter refers to several test code fragments for the programmer. These are listed in Appendix 1, with selected portions of the programmer schematic.

Cody thought a moment, then scribbled the following list:

Programmer Hardware Components

80188 Microprocessor

16550 UART/RS-232 interface

PIO chip (8255)

Control register

RAM memory

ROM memory

Vpp and Vcc DAC

"You will probably want to test this from the bottom up," Matt said. "Start with the processor and associated components, then test the memory because you'll need that for more sophisticated tests. Then verify the UART so you can use your host computer for debugging if you want, then, finally, the PIO, DACs, and control register that control your target EPROM. If you had an emulator, you could do some of this directly through it, but we'll have to do this the hard way. Come back to my office. I think I've got some code fragments from an earlier project that you can use."

Matt took the same vinyl binder from his bookcase and thumbed through it. "Here's a simple code fragment that just sets up the 80188 internal registers and then hangs in a tight loop. It will let you check out basic EPROM and RAM addressing, and let you make sure you've got the basic 80188 setup right." He turned the page around so Cody could read it.

"What do I need to use this?"

"Just a 'scope. If everything works, you'll see periodic low-going pulses on the diagnostic output (Pin 7 of the 74AC138) and on the RAM chip select signal. That will tell you that you've got the internal chip-select and other registers set up correctly."

Cody nodded. "This shouldn't take long," he said.

An hour later, Matt found Cody scratching his head over the prototype board. "Still dead?" Matt asked.

"Yes. I can't get it to do anything."

"Is the processor clock running?"

In answer, Cody touched the 'scope probe to the CLKOUT signal from the processor where it connected to the UART at pin 16. A clean square wave appeared on the screen.

"Have you checked the supply voltage?" Matt asked.

"Yes. It's in spec."

"Hmm." Matt opened the toolbox he was carrying and pulled out a logic probe. "Connect this to the processor READ line," he said.

Cody studied the schematic for a moment, then used a short clip-lead to connect the logic probe to pin 21 of the UART IC. The logic probe LED remained dark. Matt picked up the board and flexed it, looking for intermittent runs, then ran his fingers over the back, and the LED flickered. "You've got a floating pin somewhere," he said. "All your parts are CMOS, and the skin resistance of my fingers is driving something to a valid state."

Cody studied the board. "Here it is," he said. "The reset resistor and capacitor weren't installed. I remember we didn't have the parts, so we ordered some and the tech said he'd put them in later. He probably forgot." Cody went to the parts drawers and selected components, then soldered them into the circuit. Now when he turned on the power supply, the logic probe LED blinked once and went out. "That's better," Cody said. "At least it does something."

Matt reached into his toolbox once again. "I thought we might need this, so I took it out of the cabinet this morning." He handed a handwired perfboard to Cody.

"What is it?"

"It's a circuit that will let us single-step the processor. You brought one of the RDY lines out to a test point, right?"

"Just like you said," Cody replied. He studied the schematic for a moment, then clipped the four wires from the perfboard into the circuit. "How does it work?"

Matt pulled a paper from his toolbox and unfolded it. "A copy for you," he said (see Figure 4.1).

"You probably recognize the momentary switch and the two 74AC00 gates as a set-reset flip-flop wired to debounce the switch," Matt said. "The rest of the circuit drives the RDY pin high for two clocks each time the pushbutton is pressed."

"How do we use it?"

"Turn your circuit on. The stepper circuit will hold the processor in a wait state so you can look at the EPROM address and data lines to see what the data is. If the first byte from the EPROM is wrong, nothing else is going to work either."

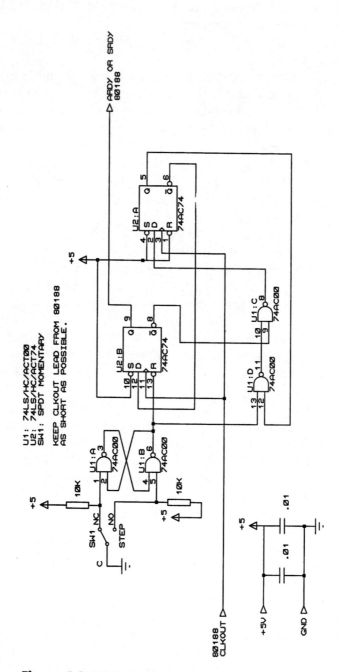

Figure 4.1 80188 Single-Step Circuit

Cody nodded and turned on the power supply, then checked each EPROM pin with the logic analyzer. He looked at this code listing. "The first byte is right," he said. He pushed the button several times, checking the data lines each time. "The data is all 1s here," he finally said.

"Look at the address lines."

Cody busied himself for a moment. "Got it. The address isn't what I expected. It's at a place where there's no code. I bet I've got a problem with the way I linked the reset code in. It's probably generating a jump to the wrong place. Let me look at this and get back to you if I still have problems."

"Don't forget to keep a log."

"Will do. Oh, and I was talking to Josh about the design. He thinks I need to be able to change the baud rate, instead of fixing it at 9600. I'd like to talk that over with you later."

"No problem."

Tests

Let's take a look at Cody's test log when he ran the rest of the tests:

Test 1 (init test). Verified that strobe and RAM CS are toggling. Verified positive and negative RS-232 charge pump voltages.

Test 2 (RAM test). Allowed test to run for 2 hours. No failures.

TxTEST1. Transmits character "A" continuously. Used 'scope to verify bit time at 104 microseconds (9600 baud). Connected to PC, started communications program, verified that continuous characters are seen by PC.

TxTEST2 (wraparound test). Connected Tx to Rx, allowed test to run 30 minutes. No errors.

PIO TEST. Verified that frequency of each succeeding pin halves.

DAC TEST. Used DSO to verify that both DACs produce sawtooth voltage.

The tests chosen for this system are typical of what you might see when debugging the hardware for a new design. Each of the test routines and the expected results are described below:

The first listing in Appendix 1 is a basic test to verify that the code sets up the 80188 internal registers correctly. The 80188 is a high integration device and can generate chip selects for the PROM, RAM, and I/O internally. These must be set up correctly before the processor can even get instructions from memory. Other processors will not have the same configuration, but may have other internal or external peripherals that must be initialized. When the code is running correctly, the 80188 will produce a regular pulse on the diagnostic output strobe (74AC138 pin 7), and a regular pulse pair on the RAM CE pin. These can be observed with a 'scope. The test also sets up the UART for 9600 baud operation, so the baud clock output can be verified as well.

The second code fragment is a RAM test, to verify the 128k SRAM. The test holds all variables in 80188 registers, so it does not need a functional RAM to run. In operation, the test writes location 0 with 0, location 2 with data 1, and so on, until the RAM is full. (The address increments by two because the code uses word writes.) After the data is written, it is verified. Then the initial value is incremented by 1 and the test runs again. If the test runs long enough, every location will be tested with every possible value. The test also toggles the bipolar LED between red and green for each read/write pass.

When running, a 'scope will reveal pulse pairs on the processor WR line (paired because of the word writes). These will appear for a few seconds, then they will go away while the data is verified. If there are no errors, the pulses will reappear every few seconds as a new set of data values are written to the RAM. If we were using an 80186 in our example, which has a 16-bit data bus, the write pulses would not be paired because only one write cycle would be needed for a word write.

Before testing the serial interface, the test log shows that the RS-232 voltages were verified. The MAX232 family of parts uses a charge pump to generate the positive and negative voltages needed to drive the RS-232 line. If the chip is not working (or is wired wrong), these voltages will not be present and the RS-232 lines will not work.

The first transmit test, TxTEST1, just continuously sends the character "A" to the serial port. This sort of test lets us check the bit timing with a 'scope and allows us to verify the slew rate of the signal. Figure 4.2 shows the waveform of one character as seen on a 'scope connected to the RS-232 transmit line while running this test. Note that the waveform is inverted (a high is the lowest voltage) with respect to the logic level prior to the RS-232 buffer. The waveform in the figure shows a single character for clarity; the actual code transmits the character continuously.

TxTEST2, the second serial test, sends a byte to the transmit half of the UART, then waits for the byte to be received and verifies it against the transmitted data. The transmitted byte is then incremented and the

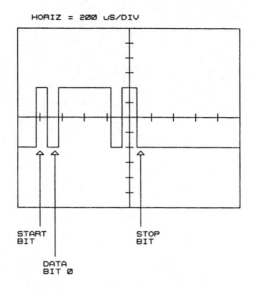

HORIZ = 200 uS/DIV

START
BIT

STOP
BIT

DATA
BIT 0

Figure 4.2 TxTEST1 Character
Waveform

process repeats, continuously sending all 256 possible 8-bit values. The bicolor LED alternates between green and red while the test is running, producing an amber color. If a transmitted byte is never received, the LED stays green, and a byte that does not verify leaves the LED red and stops the test. To execute the test, the Tx and Rx lines must be shorted together with a jumper wire (pins 2 and 3 of the DB-9 connector).

PIOTEST continuously increments a 16-bit value and writes it to ports A and B (A is the LS byte). Using a 'scope to view each port output, and working from Port A bit 0 toward Port B bit 15, each pin in turn will display a waveform with a frequency that is half that of the previous pin. This permits a single test to be used to verify PIO operation and correct connection to the PROM programming connector.

DACTEST, the last test listed, writes incrementing values to both DACs. Since the value wraps from 255 to 0, the resulting waveform is a slow sawtooth (Figure 4.3). If the 'scope sweep speed is set slow enough, the individual steps will be visible as shown in the magnified section (they won't look that clean and noise-free, of course!).

These tests are unique to this circuit, of course, but they indicate the sort of thing that can be accomplished with limited equipment and some simple software.

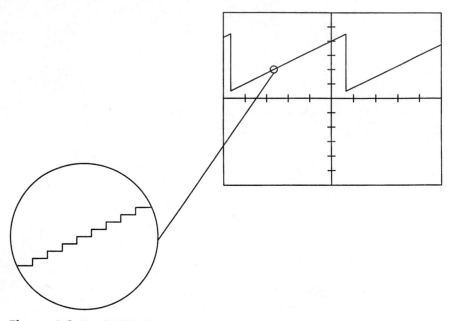

Figure 4.3 DACTEST Output

The more powerful your tools, the simpler this process can be. If you are using an emulator or software debugger, for example, the UART registers can be set up manually by directly writing to the register addresses and verifying that the baud clock is correct, the test character is correctly sent and received, and so on. One advantage to actually writing test code, however, is that the initialization code, once debugged, may be useful directly in the application code.

Hardware Problems and Solutions

Now that we've looked at the programmer to see a typical example of the hardware debug process, we'll take a look at some common hardware problems and how to go about solving them.

Floating Pins

Many years ago, an engineer came looking for suggestions about a problem he was having. He had a microprocessor circuit, with a UV-erasable EPROM, and the circuit would work only when he opened the cover of the box it was installed in, or if he put a flashlight in the box with the cover closed. It turned out the Vpp pin (where the programming voltage is applied during programming) was floating. Apparently, the chip needed just a little light (through the erasure window, which wasn't covered) to bias everything up so it would work.

In the days when everything was TTL, a floating input would show up on a 'scope as about 1 to 2 volts. Now that nearly everything is CMOS, floating inputs usually look like ground. Often, if you run your fingers over the board, circuit operation will change. Also, if you have an IC, such as an 8-bit register, that fails only with certain data patterns, look for a missing ground. Many CMOS parts will work without a ground connection as long as one of the inputs is low, but as soon as they all go high, everything stops.

It is often safe to leave unconnected, unused pins floating on a microprocessor, but I like to pull them to an inactive state, in case they are needed. Microcontrollers with internal pull-ups, of course, need no other termination in most cases.

Risetime Problems

I worked on a problem once with a circuit that would fail powerup reset intermittently, and then only on some boards. Figure 4.4 shows what was happening. Apparently to save power, the designer had used large-value pull-ups (>100k). The RAM was backed up by a battery, and the tristate buffers prevented inadvertent writes to the RAM chips during the unstable interval while power was coming up. The problem was that the 74AC08 inputs on some of the boards saw the reset go inactive before the tristate buffer did. The result was that the processor would come out of reset before the RAM circuit was ready, so it couldn't access the RAMs. Instant crash.

In another instance, a designer had used a 68000-family part with a pull-up that was too large, making the risetime of a signal longer than the specification allowed on the microprocessor data sheet. The circuit worked in production for several months, and then purchasing bought a different brand of processor chips, which was less forgiving of the input transition time.

Peripheral Timing Problems

Look at the diagram in Figure 4.5, which shows the timing cycle for a microprocessor accessing a peripheral IC. There are six timing parameters shown here, any of which can cause problems.

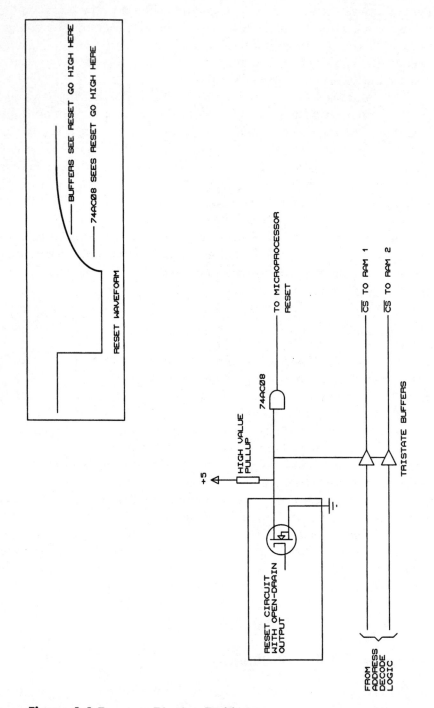

Figure 4.4 Powerup Risetime Problem

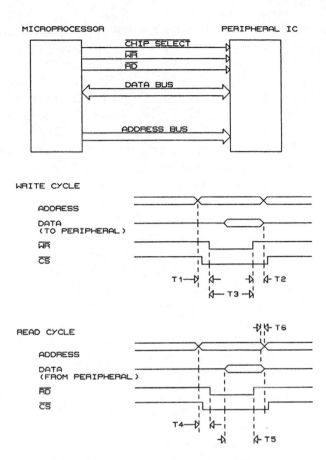

Figure 4.5 Microprocessor/Peripheral Timing

Figure 4.5 shows the typical timing for an Intel-style processor communicating with some generic peripheral IC. The chip select may come from an address decoder internal to the microprocessor itself, or from external address decode logic.

In the write cycle, the processor asserts the address, then asserts the –WR signal, then presents data to the peripheral. Time T1 on the diagram is the address setup time prior to –WR going low. If this parameter is not met, the wrong register (or memory location) inside the peripheral IC may be changed. Time T2 is the hold time of the data after the rising edge of –WR. If this is not met, the peripheral may store the wrong data. The last parameter, T3, is the minimum length of the –WR pulse itself. Some peripherals have a maximum parameter as well.

In the read cycle, time T4 is the address setup time prior to the falling edge of –RD. Not meeting this parameter is usually less critical than the equivalent parameter in the write cycle, unless the peripheral latches the address on the falling edge of –RD. Time T5 is the time the data must be stable prior to the rising edge of –RD and is effectively the access time of the peripheral IC. If it is not met, the processor may read the wrong data. Time T6 is the hold time of the data after the read cycle is complete. This is more likely to be a problem on a processor with a multiplexed address/data bus, where a peripheral that does not release the bus quickly enough can cause bus contention on the next cycle.

These parameters are typical of what you see in processor/peripheral data sheets. There are others, of course. Some peripherals have a parameter on the minimum time between successive accesses or require that input signals be synchronized to a clock. Sometimes they want write data to be stable before the leading edge of –WR, which requires additional logic on Intel-type processors. Other processors, such as the Motorola 68000 family, have different cycle and signal structures, but the same types of timing requirements apply.

Many designers just connect peripherals together, assuming that if the clock rates or the access times are right, everything else will work too. This can be dangerous, especially if production will run for many months or years, giving plenty of opportunity for parts from different manufacturing lots to be installed. Finding a timing problem is best done when the design is started, since fixing one can add a significant amount of logic to the board. You should verify that all timing parameters are met.

Risetime problems, timing problems, and floating-pin problems will often be temperature sensitive, showing up only when hot or cold. This is because the thresholds and speeds of the parts shift slightly with temperature. If you have an intermittent problem that you suspect is caused by one of these conditions, you can often make it show up by using circuit cooling spray or a hair drier to cool or heat the board. Be careful you don't get it so cold the plastic IC packages crack, or so hot the packages melt.

EMI Problems

Embedded systems often must control stepper motors, DC motors, or relays. All of these can cause electromagnetic interference (EMI) problems. Any inductive device will cause EMI when it is switched on and off. Whether the EMI causes problems is another matter.

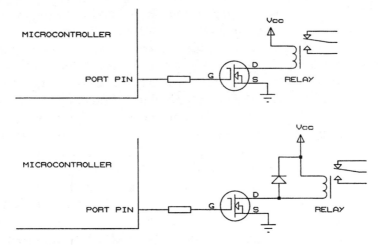

Figure 4.6 Microcontroller Driving a Relay

Figure 4.6 shows a microcontroller driving a relay through a port pin and a MOSFET transistor. In the top diagram, there is no protection diode between the transistor drain and the supply. When the port pin goes low, the MOSFET will turn off, opening the relay. However, the energy stored in the relay has to go somewhere; the result is a massive voltage spike on the drain of the transistor. Depending on the characteristics of the relay coil and the transistor, this flyback voltage can approach 100v—enough to destroy the transistor. The solution, shown in the lower diagram, is to add a snubber diode across the relay coil. The transistor drain is now clamped to one diode drop above the positive supply. For faster opening of the relay, a transient suppresser diode can be used instead, allowing the drain voltage to rise to some voltage between the supply and total destruction. If using a transient suppresser, remember that the drain voltage will rise to the *sum* of the supply voltage and the transient suppresser clamp voltage.

The catch, which is often overlooked, is that this fix is not really free. Adding a diode protects the transistor, but the coil energy still has to go somewhere, and it does. It takes the form of a current spike into the positive supply. If the supply is poorly bypassed, the result may be a voltage spike on the supply itself. When driving relays (or DC motors, or solenoids), take a little extra care to be sure the supply has adequate bypassing and that the path between the relay and the supply has a low impedance.

Figure 4.7 shows a microprocessor-based board driving a motor. For our purposes, the motor could be a DC motor, a stepper, or almost anything. When the motor is turned on, there will be a current increase, which passes through the ground wires back to the power supply and to chassis. The current causes a voltage drop, as indicated in the figure, because the wiring impedance (inductance plus resistance) is never zero. If the wiring inductance is high enough, the ground on the processor board can be upset far enough to affect operation or to corrupt communication with other boards in the system. This can be a particular problem with stepper motors or with DC motors that are PWM (pulse-width modulation) controlled, since there is usually a high-frequency surge every time the current is turned on.

Figure 4.8 shows a way to minimize this problem. A third ground wire, not connected to logic ground, has been added to the system, and returns motor current to the power supply. The motor still causes a current surge, but it does not affect the logic ground. This solves the EMI problem, but it can cause other problems: the motor is usually driven from an H-bridge or other driver, and if the motor return and logic ground get too far apart, the H-bridge may be damaged.

Ground Loops

The classic case of a ground loop is two circuits that are connected to different grounds, and to each other, but the grounds have slightly different AC or DC potentials. Since the impedance between the

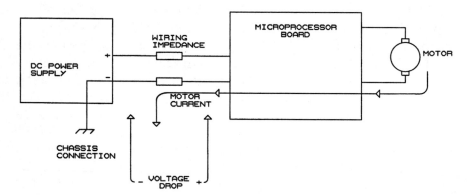

Figure 4.7 Motor EMI Problem

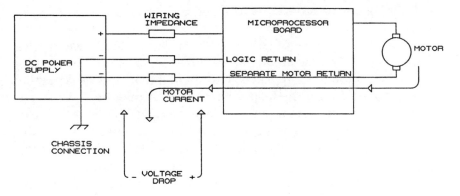

Figure 4.8 Fixing the Motor EMI Problem

two grounds is very low, significant current can flow in the grounds themselves.

Figure 4.9 shows an embedded microprocessor system communicating with a host PC. Both are powered by the 115v AC line. If the two systems are connected to different branches of the AC circuit (say, if they are in different rooms or different buildings), there can be significant current flowing in the ground. This current flows through the ground wires in the interface connections.

The problem of ground loop can be particularly bad if the two systems operate on different AC voltages. For example, if the microprocessor system is part of a large machine that requires 208V 3-phase power, real problems can occur. I have seen RS-422 drivers literally destroyed when the ground of an embedded system got yanked around by air

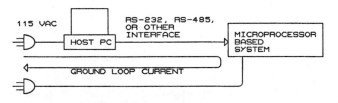

Figure 4.9 AC Ground Loop

conditioning compressors (which shared the 3-phase power) going on and off.

If the interface between our host PC and embedded system is RS-232 or serial RS-422, the problem can sometimes be solved by running the interface through an optical isolator pair. If the interface is parallel, a LAN, or some other high-speed interface, it may be necessary to ensure that the two systems are on the same branch of the AC line. In the case where the voltages are different, you may have to be sure that they both have clean, independent returns to the building ground, with no heavy-duty equipment sharing the ground return.

This same problem can occur within an embedded system that is composed of many boards and modules, each of which has a separate power supply. Sometimes you can fix these problems with a ferrite bead on the right cable, but that tends not to be a very permanent or repeatable fix.

Low-Level Signals

Ground loops do not have to affect the processor directly. They can instead affect the devices to which the processor connects. Figure 4.10 shows a processor board using a thermistor to read temperature. The thermistor has a fairly low output level—say, 1 mV per degree. The logic on the board where the thermistor is draws current, which causes voltage drops across the power supply wiring and all the connectors. The normal voltage drop typical of such systems is not enough to upset the logic, but it can be sufficient to cause an offset in the thermistor reading. Worse, the value may change as the DC current changes with the state of the logic. The solution to this problem (which was used in a similar real-world situation) was to give the thermistor a separate return so it was not affected by the offset voltage (see Figure 4.11). Of course, the same principle applies to strain gauges, pressure transducers, or any other low-level analog input device.

Shorted Outputs

Another source of EMI problems is shorted outputs. It has been my experience that having two CMOS or TTL outputs shorted together makes the overall circuit susceptible to noise, and, of course, the shorted outputs themselves dump a lot of noise into the grounds.

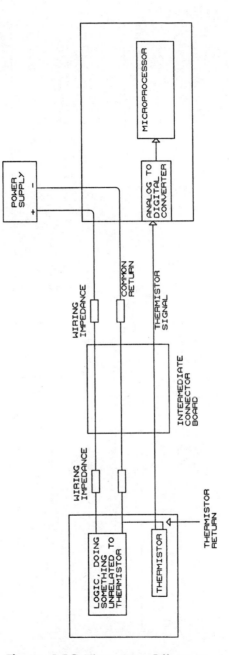

Figure 4.10 Thermistor Offset

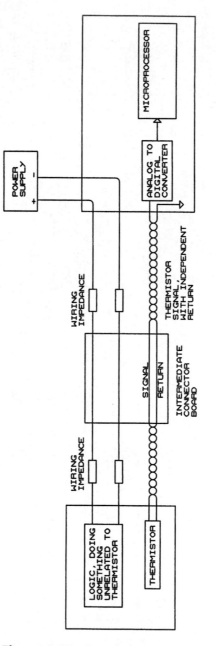

Figure 4.11 Thermistor Circuit
with Separate Thermistor Ground

Self-Generated ESD

ESD (electrostatic discharge) will often cause upset in a microprocessor-based circuit. The ESD pulse contains very high frequencies that couple very readily into the logic. Some equipment sold in Europe must be tested for resistance to ESD. However, it is also possible for electrome-chanical equipment to generate ESD internally. If your system uses rotating motors, especially if the motors are coupled to a pulley using a belt, and if you have bizarre failures, look for ESD. A motor driving a pulley with a belt made of insulating material can be a good generator of ESD. This is also true for two insulators rubbing against each other, such as a plastic brake used to prevent coasting on a plas-tic drum.

The usual solution to ESD is to use belts and pulleys that are slightly conductive. If this is not possible, you may have to use a conductive brush to carry the charge to ground, or you may need to look at alternative drive mechanisms.

The Problem Went Away When I Hooked Up the 'Scope

This happens far too often. A subtle bug causes you to tear your hair out, but when you hook up the logic analyzer or 'scope to look at it, it goes away. When this happens, look for timing errors or race conditions. Usually, the test equipment is adding a few pf of capacitance, enough to slow down the risetime of some signal.

Race Conditions

This topic is more properly suited to a book on logic design, but it often pops up in embedded systems.

Figure 4.12 shows a 74AC139 driven by a microcontroller to generate pulses to some external system. Using the 139 allows the controller to generate four separate outputs using only two port lines. The outputs could generate interrupts to other boards, or could clock data into registers, or something similar.

The timing diagram in the figure shows what happens as the microcontroller steps through the select lines. As each input line changes state, there is a momentary glitch at one or more outputs. The diagram shows glitches on the Y1 and Y3 outputs; this could vary from one manufacturer to another, based on the internal structure that each uses. Of course, if these outputs are driving registers or latches that are

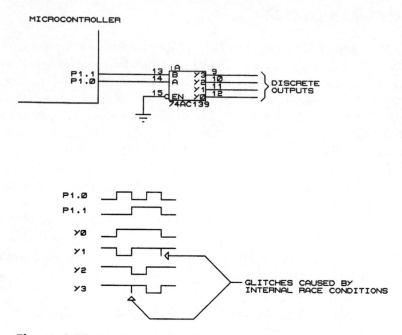

Figure 4.12 Race Condition

fast enough to respond to the glitches, the result could be invalid data in the registers. The solution to this particular problem is to use a third port pin, connected to the enable of the AC139, to gate the outputs off when the select inputs are being changed.

In Chapter 5, we'll look at software debug.

5

Debugging the Software

"I've got the hardware checked out," Cody said. "Any idea how I can handle the baud rate selection without adding hardware?"

"Let's look at the schematic," Matt suggested.

Cody placed the drawings on the desk. "I've used all three of the 8255 ports for the PROM in the programming socket. I've got four bits that I can read from the programming adapter, but I don't want to use those."

"What baud rates do you need?"

"Josh said 9600, 19200, and 38400 would be enough."

"Can you use jumpers or a DIP switch inside the box?"

"Sure. We don't expect to change the baud rate once a system is set up and running. He just doesn't want to make downloads unnecessarily slow if the programmer is connected to a computer that can run faster."

"Makes sense. A 64k download will take about a minute at 9600 baud, about 17 seconds at 38400." Matt looked at the schematic. "It looks like you've got three unused interrupts here," he said. "INT0 is used for the UART, but INT1 through INT3 are free. What if you add shunt jumpers from two of the interrupt pins to ground, and then pull the interrupt pins up with resistors? Then you've got two bits you can use." Matt sketched on Cody's schematic (see Figure 5.1).

"How do I read them?" Cody asked.

"Well, the hard way is to set up a vector for each interrupt and unmask it once after powerup. The ISRs each set a flag, then mask the interrupt before returning. You enable interrupts, wait a few cycles, then check the two flags to see which of the interrupts occurred. That tells you which pins are high and which are grounded by the jumpers. On the 80188, though, there's an easier way. You set the interrupts to be

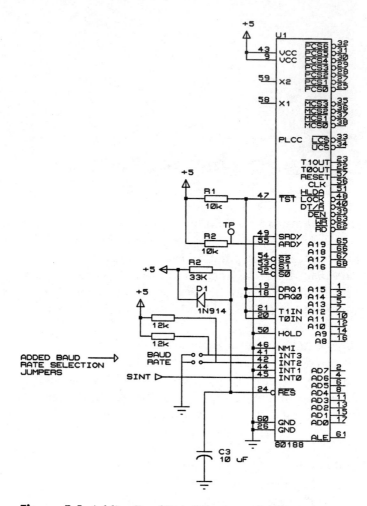

Figure 5.1 Adding Baud Rate Selection to the Programmer

level-sensitive and read them. The 80188 has a register that lets you read the state of an interrupt pin even if it's masked."

"That sounds okay. I've got a small breadboard area on the board where I can put the jumpers."

"Do you know that you've got the throughput to support the higher baud rates?"

"Not really. I don't even have the code working yet."

"We'll need to check that later."

Overview of Programmer Code

While Cody adds the baud-rate selection jumpers and the associated code to the programmer, we'll look at selected log entries he makes while debugging the code. First, though, a brief explanation of how the programmer code works will allow all this to make sense.

Figure 5.2 shows the major routines used by the programmer software. The bar at the top is the background (or polling) loop. It consists of three functions:

1. Rx FIFO Check looks for data in the receive FIFO (determined by the read and write pointers being unequal). If there is data, RXPROCESS is called and checks to see whether a data value, a command sequence, or a command parameter is expected. RXPROCESS then calls the appropriate routine to process the command, data, or command parameter.
2. Pushbutton Check looks for and debounces the pushbutton. This code calls nothing but sets appropriate flags, depending on the current mode of operation.
3. Device Mode Check does nothing if the programmer is not programming, blank checking, or verifying a PROM. However, if a device operation is in progress, then the appropriate device code is called—PROGRAM, VERIFY, or BLANKC. These routines in turn call lower-level routines that perform an operation on a single location in the target PROM.

The original background loop would have included a Tx process routine for transmitting data to the host, but this was removed, as will be discussed in a later chapter.

Various support routines are provided as well. The message output code, called from many locations, is called with a pointer to the message to be sent and just stuffs data into the transmit FIFO until an end-of-message is detected.

The device table support routines return various parameters, such as the device size, blank value, programming pulse width, and program/read voltages. The tables are indexed with the current device type.

HEXASC and ASCHEX convert hex data to ASCII and ASCII data to hex. They both operate on global variables, HEX and ASC. HEX is a 2-byte variable, and ASC is a 4-byte variable. Both routines convert the complete contents of the input buffer (HEX for HEXASC and ASC for ASCHEX) regardless of how many bytes the caller actually is using.

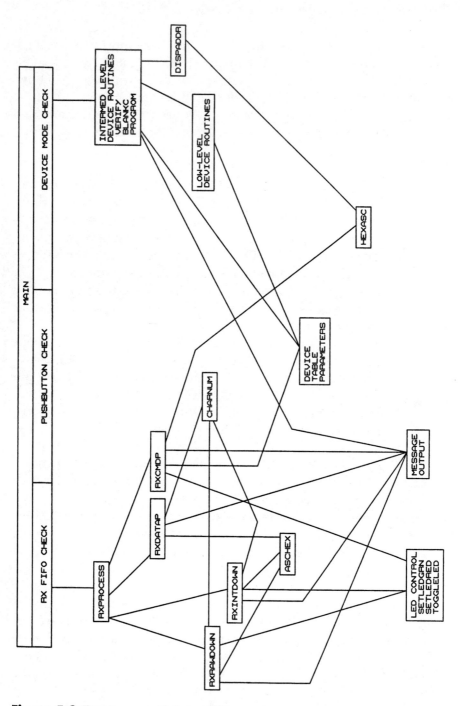

Figure 5.2 Programmer Software Chart

Now that we have a limited overview of the code, we'll look at some of the problems that were encountered:

Programmer Software Debug Log

Want to get debug commands (#XM, #XB) working first so they can be used for everything else.

#XM display should look like this (one line):

aaaa: dd dd dd dd dd dd dd dd dd dd dd dd dd dd dd dd dd

aaaa = address, dd = data, display repeats until 64 bytes displayed. Attempting to display address of 0 results in display of 000D first time. ASCII-to-hex conversion assumes data is checked for validity, but nybble conversions use only first byte of the global ASCII variable. Data in second byte is expected to be legal ASCII by conversion program, but was uninitialized and came up with value larger than 80h. Since it wasn't used by the first caller, it carried into first byte result. Fixed by initializing ASC[0-3] to 0 in initialization code.

Binary download: Data downloaded always came out with data in the wrong place. Created short file, it worked okay. Used #XB to display buffer contents after download and compared it with original hex file (which was converted to binary to test download). Some data was missing—all bytes with value of 0Ah. The short test file did not contain any bytes of 0Ah. Communication program was stripping 0A (linefeed) out of file while sending. No fix needed, but this will be a problem if programmer is controlled by off-the-shelf communications program and binary download is used.

Tried to perform blank check function on empty socket. Processor resets. Cleared trace buffer after powerup and tried it again, found second reset action code in trace buffer, so the processor is really going through the reset vector. Code hierarchy for this command looks like this:

Blank check in background code

> *BLANKC routine to blank check a byte*
> *Device Support routines to get device size, blank value, etc.*

The blank check action code is placed in the trace buffer on the first pass through BLANKC, so that part works okay, and the command is being decoded properly. Added action codes of 80h, 81h, 82h to BLANKC just before first call to device support routines, to first device support routine that BLANKC uses (READSET), and to BLANKC after return from that first call. The 81 gets put in the trace buffer, but the rest of the codes don't.

Found problem: Device support routines get a value that determines which of the program/read/setup routines to use, based on device type. This value is used as an index into a jump table that directs execution to the right routine. But the vectors are word length and the index wasn't being doubled to account for that.

When programming, verify fails. Examining bad PROM in commercial programmer reveals that ones are changing to zeros at the bad locations. The problem therefore is not insufficient programming pulse width. When programming, the algorithm programs a byte, then verifies it to determine whether programming is done. Disabled verify after program and changed program code to just program each location 10 times without checking. PROMs now OK, so problem is in verify code.

Found bug in code: The code that verifies each byte after programming (i.e., not the code that verifies the entire device after programming complete) was doing this:

> *Float the data lines*
> *Drive CE low*
> *Drive OE low*
> *Read & verify the data*

Since CE low, OE high, and Vpp at program level causes the PROM to program, this sequence was causing a brief programming pulse with the data lines floating. Swapped the code so that OE is driven low prior to CE, keeping the device from going into programming state when it should be verifying.

When Cody debugged this code, he had very limited tools. Let's look at how these problems might have been debugged if we were using different tools.

Uninitialized Variable

This problem could be solved with an emulator or ROM monitor by using the following sequence:

Run the code until entry of HEX to ASCII conversion code.

Single step through the code, examining the ASCII output data until the bad byte is seen.

Look backward in the trace buffer for the math operation that caused the problem. This will lead to the bad data in the uninitialized location.

Binary Download

This is a problem with the development environment. There are not a lot of really fast ways to find this one. You could create and download a short file that contains every possible value from zero to 255, but a binary search through the buffer for the bad data is probably just as fast.

Incorrect Vector Index Causing Reset

The process of adding trace outputs required recompiling the code and burning an EPROM (or downloading, if a PROM emulator is being used). If a logic analyzer was being used to capture trace data output to a diagnostic address, the effect would be the same. Using an emulator or ROM monitor:

Set a breakpoint at the first call to BLANKC, then single step through the code until the problem appears.

Or, set breakpoints at each of the locations that action codes were added, then run until a breakpoint is reached. The breakpoints can be moved up until the problem is isolated to the vector branch.

Of course, if the programmer software were written in an HLL, this problem would not have appeared, at least in this form. However, as any C programmer can attest, there are a number of other pointer problems that can occur.

Bad Verify Sequence Altering PROMs

This was a hardware interaction problem. It is not possible to trigger an analyzer on the bad data, since the problem will occur at some locations and not at others. If an emulator or ROM monitor is being used, a breakpoint can be set when the verify code fails, but that will not help in determining what the actual problem is. Probably the best method of finding this one would be to use a logic analyzer to look at the address, data, and control lines of the target device while programming. The incorrect sequence will be more obvious on the displayed waveform than in the code listing.

Modular Testing

There are two camps in the issue of software testing. One position is that you should test and debug the entire program as a unit. The other extreme holds that you should test each module or routine before integrating them together into a single program. The tradeoff is that testing each module ensures a final program that is more likely to work, but the testing itself can take significant time—time that is wasted on those modules that work the first time. The test time issue is particularly significant if the data to test the module is difficult or time-consuming to create. If the module interfaces to a piece of hardware, it may be difficult to create conditions that result in a meaningful test if the rest of the hardware and software isn't running.

Testing modules is most effective on a simulator or using a debugger on the target. It is usually just too tedious to do this testing without some means of starting and stopping the program at will and looking at the results directly. This rules out action code outputs and memory dumps for most cases.

Of course, as for most things in life, there is a position between the two extremes. I like to test modules that are risky (because of execution time), or that are straightforward to test. The following programmer modules illustrate the issues involved in this type of testing:

ASCHEX and HEXASC, which convert between ASCII and hex, would be fairly easy to test independently. Using a ROM monitor or emulator, you would place data into the appropriate buffer, execute the routine, and verify the result (again, using the debugger tool).

The serial subroutines present a different problem. The receive ISR, for example, places received data into the Rx FIFO, where it is processed later. The following description shows the receive ISR functionality (RWPOINT is the write pointer for the Rx FIFO):

1. Save registers on stack.
2. Read data from UART.
3. Store received data at RXFIFO[RWPOINT].
4. Increment RWPOINT.
5. If RWPOINT wraps past end of buffer, set to beginning of buffer.
6. Send EOI to interrupt controller.

7. Restore registers from stack.
8. Reenable interrupts.
9. Return.

The functionality of the FIFO operations could be tested using a debugger by placing data in the appropriate register and then running the code from statements 3 through 5, then examining the FIFO and RWPOINT. However, this does not verify the interaction of the code with the UART and the interrupt controller. For these, the UART must get real data and generate a real interrupt. This means that there must be something on the other end of the UART to generate data for the test.

For an even more extreme example, consider a motor that uses a trapezoidal speed profile to move some mechanical assembly back and forth on a threaded shaft. The motor is geared so that it requires several revolutions of the motor for one revolution of the shaft. The shaft encoder produces, say, one pulse per revolution of the threaded shaft.

To move the assembly to a new position, the motor ramps up to a constant speed, holds that speed for some time, then ramps back down to a stop. Figure 5.3 shows the motor speed waveform, the encoder pulses, and a mechanical schematic of the mechanism.

Say that there are three routines for this move, one for ramp-up, one for constant speed, and one for ramp-down. They all control the speed based on input pulses from a shaft encoder. Testing the constant-speed module requires that the motor and shaft assembly actually be connected, or that a very good simulation of the motor/encoder pair be used. Such a simulation would have to respond to motor speed commands the same way that the actual mechanical assembly would, complete with frictional and inertia effects. Such a simulation would be difficult and time-consuming to write, and at least as difficult to verify as the module we want to test.

On the other hand, if the actual motor/encoder is used, we have to have the ramp-up module installed and working to get the motor to the correct speed. Then, we must have some way to verify that the constant-speed module works. Since the output is the motor speed profile itself, this will be difficult to do with only software tools. Last, we need the ramp-down routine working so we can stop the motor at the end of the move. If we get halfway through the constant speed routine and decide to stop execution so we can look at some variables, the processor will miss several encoder pulses, leaving us with no idea what the final position is. In addition, if there isn't some provision to turn off drive to the motor, it will keep going until it spins the mechanical assembly to the end of the shaft.

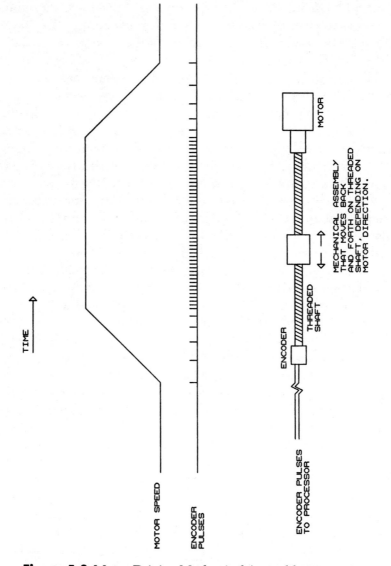

Figure 5.3 Motor-Driving Mechanical Assembly on Threaded Shaft. Note trapezoidal motor velocity profile.

Clearly, this motor application is one where module-level testing presents numerous problems. In general, any code that has to deal with hardware that will not just sit there while you look at things with the debugger is usually a poor candidate for module-level testing. The more external hardware or special test code you have to write to independently exercise a module, the less attractive it is to do so.

Regardless of whether you test individual modules or bring up all the code at once, you will reach a point where you have eliminated the obvious errors so that the code doesn't crash every time you start it, and everything appears to be working. This is the point where functional testing begins.

Functional Testing

As the name implies, this is the level of test where you verify that the program actually does what it was intended to do. What a functional test looks like depends on the complexity of the system to be tested. In general, a functional test should include the following:

Test all functions for normal operation. You'd be surprised how often something is not tested because it's a tedious test to do and the engineer is just "sure it's right."

Test all the extremes and rollover conditions. What does the code do at the end of a minute/hour/day/year/century?

Test all the special cases: leap year, entry started but aborted by operator, anything where the code takes a path different than the one for normal operation.

Test all error conditions that the software checks for.

Test all the user mistakes you can think of. What if the operator holds down two keys while pressing a third? What if she enters the date as April 31?

Test with all the inputs running. You will probably have to plug into a real machine and run in the actual environment. On the other hand, if the real machine is 5,000 miles away and everything must be simulated so someone else can install it there, make sure your simulation really represents the real world.

As an example, the functional test for the programmer might look like this:

Verify blank check on a blank and nonblank device of each type.

Verify that device offset works by programming a PROM using offset.

Verify that buffer offset works by programming a PROM using offset. Verify with Intel-format and raw hex files.

Verify that device size function works for programming, verifying, and blank check functions.

Verify that Intel-format file loads correctly.

Verify that extended Intel-format loads correctly.

Create a file that causes address wraparound using extended Intel, and verify correct operation.

Verify error functionality by loading a file with one byte that is different from a programmed PROM. Verify that error is captured and correct address is displayed.

Verify a PROM programmed from Intel-format hex against raw hex and vice-versa.

Verify that a raw hex file greater than 64k does not crash software.

Verify that Intel-format hex file with out-of-sequence addresses works correctly.

Verify that Intel-format hex code correctly detects bad checksums.

Verify handling of invalid input command strings.

Verify that #Q and pushbutton stop the commands they are intended to affect.

In theory, a functional test will cover all the extremes and all the special cases. In practice, some of these are always overlooked until they turn out to be a problem, after which they are added to future functional test plans. This is one of those times when theory runs into the brick wall of reality.

Debugging

Even with all the module tests, at some point functional testing will turn up a bug. This is where testing becomes debugging.

Your Arsenal

If you are using a ROM monitor or a hardware emulator, your primary debug tool will usually be the breakpoint. If you are using action codes in a trace buffer or to a debug port, then these will be your tools.

Types of Software Errors

For our purposes, we will categorize software errors into the following basic flavors:

> Design errors: Misunderstanding the requirements or having the requirements change. "Oh, I have to control the pump, too?"
>
> Specification conflicts: "Part of process 3A requires having the pump on with the door closed, but the specification says the pump must always be off when the door is closed."
>
> Logic errors: I wanted to do this, but I coded that. "The way I coded it, the code doesn't check to see if the door is open before turning the pump on."
>
> Coding errors: "Oops. I wanted to turn the pump on when the door was open, but I turned it on when the door was closed." This category includes syntactical errors, such as misspelled variable names.
>
> Uninitialized variables: Self-explanatory.
>
> Timing errors: "I turned the pump on before the door was fully open." This category includes race conditions.
>
> Synchronization errors: "This function thought the system was in one state, but that function thought it was in another state."
>
> Missed deadlines: "I turned the pump off, but not soon enough."

Of course, these are not exclusive lists—an uninitialized variable can cause, for instance, a timing problem, or a timing problem could cause a synchronization error. And some problems may fit into more than one category.

Problems involving a misunderstanding of the requirements are usually easy to resolve—the code has to change to meet the requirements. The problem is more difficult if, for example, the hardware was based on estimations of the required software throughput. In such a case you have to change the hardware or the requirements.

An example of this scenario involved a microprocessor-based subsystem that I worked on. We had decided to use existing processor and interface hardware that was already used in other parts of the system. The software engineer elected to use an object-oriented language, our first entry into that particular quagmire.

The first difficulty came about when we discovered that the compiler was incapable of producing code that would run in the available memory. We had to add memory, both EPROM and RAM, and this meant changing the PLDs that mapped the memory. The processor was from the Intel x86 family, so it had an I/O space separate from the memory space. To make the hardware fit the object-oriented architecture that was selected, all the I/O devices had to be mapped into the memory space, requiring more PLD changes. Finally, we found that the code couldn't execute fast enough to meet the timing requirements, so we had to go to a faster version of the processor and change the crystal (the rest of the circuitry was already designed to accommodate a higher clock rate).

You could argue, of course, that this was a serious misapplication of the language chosen. But by the time we discovered all the problems, there wasn't time to rewrite the code in a different language. So in addition to code that could just barely keep up with the rest of the system, we wound up with unique hardware for that subsystem, which meant extra parts in inventory and more part numbers to support.

Requirements that change in the middle of a project is a topic that is really outside the scope of this book, although it often indicates that someone failed to do his job up front, or that the design started too early.

Specification conflicts are usually fairly obvious, although sometimes not until coding begins. The more difficult situation is specification conflicts that are not stated in a conflicting way, but are implied by two different and supposedly independent requirements. The conflict may not even be a conflict except under certain circumstances and when the code is written with certain assumptions. These types of conflicts may not show up as requirements that are impossible to code, but as something that forces the software into an invalid or ambiguous state.

As an example, let's look at the motor/screw assembly from Figure 5.3. Let's say that one requirement is that the speed be below 500 revs/minute if the mechanical assembly is within 3 inches of either end of the threaded shaft. This is to prevent resonance and vibration problems from the long, unsupported shaft. Our system also has a second requirement that all moves be completed in under 5 seconds. Our hypothetical code to control the motor is structured like this:

At the start of a move, the code creates a velocity profile, based on the distance to move and the starting position. The motor is controlled by a PWM mechanism, where the processor sets the duty cycle of an internal timer, which drives switching transistors that turn the motor on and off.

Each time a pulse is received from the encoder, the code calculates the time that should have elapsed since the last encoder pulse. If the actual time is greater than the calculated time, motor speed is too slow, and the motor drive is increased. If the actual time is less than the calculated value, the motor speed is too high and the motor drive is reduced. If the motor speed is over the 500-rpm limit while the mechanical assembly is within the 3-inch limit, the code turns off motor drive, allowing the motor to coast, until the speed is under 500 rpm.

Figure 5.4A shows a move from about 6.5 inches (from one end of the shaft) to about 2.5 inches. The processor selects a profile, as shown, and begins the move. However, as shown in part B of the figure, when the motor reaches the 3-inch point, the speed is over 500 rpm, so the motor drive is turned off and the motor coasts until the speed is below 500 rpm. Then the move is completed, but the angle of the slowdown ramp is different than planned.

The overall slowdown ramp is very similar to the intended ramp, but takes slightly longer. The point where the mechanism stops is the same, so everything appears to be in order. Of course, the fix for this is to take the final stopping point into account when creating the initial profile, ending up with a profile that has shallower ramps but a higher constant speed. That's once you find the problem. But this example illustrates some important issues that can result from two conflicting requirements. Possible results of this bug are as follows:

1. Some moves (but only some) take slightly longer than the 5-second maximum, causing you to think that the code is incorrectly predicting how long a calculated velocity profile will take.
2. Vibration/resonance problems may be encountered since the speed is over 500 rpm at the 3-inch limit. But, since the code corrects the condition, the speed is not over the limit for very long, so the result may be subtle problems that are attributed to mechanical tolerances.
3. The (relatively) sudden changes in velocity may introduce unexpected problems in whatever the mechanical assembly is supposed to do, or additional resonance problems that were not initially taken into account. This situation would be

made worse if the motor were braked to bring the speed within limits, instead of being allowed to coast.

Finding this particular problem would involve something like the following process: First, there would be an Aha! step where you determine that the motor is coasting because of the limit violation. If you suspect this to be the case, proving it is happening could be as simple as producing a trace output or setting a breakpoint in the part of the code that detects the 500 rpm/3-inch condition. Another way to detect this is to look at the PWM drive with a logic analyzer or DSO, set to trigger if more than x milliseconds go by with no drive pulses.

If you don't start with the assumption that the motor is coasting, you might connect a logic analyzer to capture the encoder pulses, measuring the time between them and comparing them with the calculated profile. This would take a while, since the profile consists of probably hundreds of steps, each of which must be compared with the actual time. Remember, there isn't a problem until near the end of the move. Very tedious, and probably not the first thing you would try.

Once you do determine that the actual profile doesn't match the intended one, you have to determine why, and you may look in the motor speed control code to see whether there is a calculation error. You might also check for missing encoder pulses, or extra pulses caused by noise.

At some point, you find what is actually happening. In our example, the cause of the problem, and the solution, is fairly obvious (but possibly difficult to implement, given that a third variable must be added to the profile calculation). In a real system the two conflicting requirements may be in two separate specs, such as a product requirements document and a design spec that details how the shaft/motor assembly is to be driven.

Another example, from a real design, goes like this. A system was designed to feed items one at a time into a conveyor system. The system had two requirements: a certain throughput in items per minute had to be maintained, and the distance between items (the spacing) had to be greater than a certain value. The spacing requirement came from a need to push the items into hoppers (downstream of the feed mechanism). The mechanical gates that directed the items into the hoppers needed a certain amount of time to get out of the way before the next item arrived.

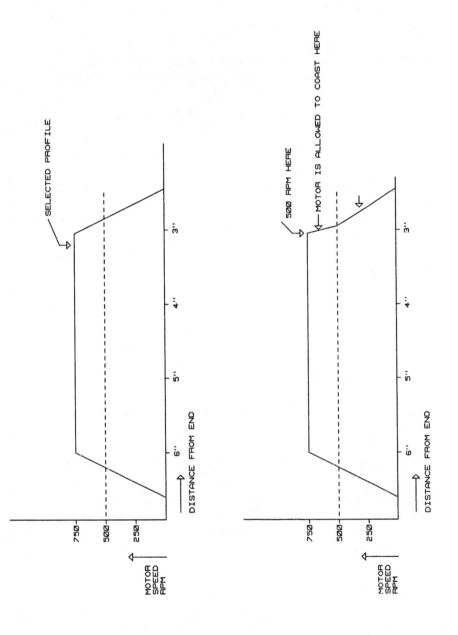

Figure 5.4 Intended Motor Velocity Profile and Actual Profile

These requirements did not appear to be in conflict, at least on the surface. The distance requirement was such that it was possible to meet the throughput requirement.

The feeding hardware consisted of a DC motor with speed control that came from a PWM driver. The processor passed a speed byte to the motor drive circuitry, and a servo loop maintained the motor at this speed.

To feed items, the software kept the motor running at a terminal speed. On a regular basis, the software would check the throughput and adjust the terminal speed up or down to maintain the correct throughput value.

For every item fed, the software would also monitor the distance between the item in the feeder and the item just fed. If it predicted, based on motor speed, that the distance between the two was going to be too small, it performed a correction by ramping the feed motor down, slowing the new item and correcting the distance, then ramping the motor back up to the terminal speed.

This arrangement worked when the hardware was well adjusted and the items were in good condition. If the distance correction function had to correct too many items, the throughput would go down, so the throughput code would compensate by raising the terminal speed.

The problem occurred when the drive components were worn or the items fed were in poor condition, which caused slippage and skewing. The throughput requirement could not be met, so the terminal speed was raised. This caused the distance correction function to correct nearly every item, lowering throughput still further. The result was that the terminal speed would ramp up to the maximum value, but the throughput was only about 75% of what was specified.

The correction for this problem was to add coupling (negative feedback) between the two functions so that the terminal speed would not be raised if there were too many or very severe corrections. The key point, though, is that the bad state occurred because two legitimate requirements were not simultaneously accounted for in the code.

Coding and Logic Errors

These errors are usually fairly straightforward to find, and we won't spend a lot of time on them. The usual process is to use breakpoints or action codes to determine about where the problem is, then the code is examined and/or traced to find the error.

Uninitialized Variables

These variables are debugged in much the same way as coding/logic problems. There are a few additional caveats for these problems, however.

An uninitialized variable is a location in RAM. The RAM can be inside the processor, as in a microcontroller, or a separate IC connected to the processor.

When power is applied, the RAM comes up in an unknown state. The value contained in the uninitialized location may or may not cause problems in the code, especially if the variable is a flag or semaphore that is tested for zero/nonzero. Since the RAM may come up one way at one time and another way another time, the problem may be intermittent, although it will go away the first time the code writes to that variable. In theory, an uninitialized 8-bit semaphore that is tested for zero/nonzero has a 1 in 256 chance of being bad (or 255 in 256, depending on which state is considered bad).

A more common scenario is for a specific location in a given RAM IC always to come up in the same state. This is especially true for static RAM. If the powerup state of the variable happens to be good (say, zero), no problem will be seen. Until, that is, the system goes into production where RAM parts from a different manufacturing lot or a different manufacturer are used. Some of these may come up with no problems while others make the system fail.

One real incident involving an uninitialized variable worked out exactly that way. A flag value came up as zero (every time) on the engineering unit that was used to test a new version of the code. In manufacturing, about half the boards came up with a bad value. Unfortunately, for certain reasons having to do with the customer's validation procedures, we couldn't change the code. Shipments were stopped and a major revenue-producing product was essentially shut down until this problem was fixed. The solution was to change the BOM (Bill of Materials) for the board to use an NVRAM instead of a normal static RAM. The NVRAM, using an internal battery, retained the RAM contents during powerdown. The factory would bring the system up, run the system in such a way as to force the code to set the bad location, and from then on, everything was OK.

Timing Errors

These errors come in various guises and are often the most difficult problems to debug. One variation involves the fact that physical processes often take time to complete, time that must be taken into account in the software.

As an example, let's look again at our motor/shaft/mechanical assembly system. Suppose the code is structured so that one routine sets a flag somewhere to request a motor move. The routine that actually moves the motor resets the flag to acknowledge that the move has started. The problem occurs if this flag is treated as a reset-means-ready-to-move-again condition. The move routine clears the flag at the beginning of the move, so the routine that requests moves requests another one.

A more subtle version of this scenario would be if the requesting routine waited 5 seconds after a move started (because that's the maximum time a move is supposed to take), then assumed that the move was complete. In our example where the speed/distance limits were exceeded, the bad move might take more than 5 seconds. So we might have a problem that shows up only when:

1. The move exceeds the speed/distance limits, and
2. The resulting move takes slightly longer than 5 seconds, and
3. The move-requesting routine wants to make another move before the bad move is completed.

Since all three conditions have to be satisfied for the bug to appear, the result may be a problem that shows up very rarely. The various paths you might go down before finally hitting on the real problem could be:

Sometimes the requester doesn't really wait 5 seconds (but why?).

The problem is position related—it shows up only when the moving assembly is near the ends of the shaft (because the limit violation only occurs when the move terminates within the 3-inch boundary).

The motor move function sometimes cannot handle back-to-back requests (but why only near the ends?).

Another variation of this problem occurs when there are two processors. One processor passes a command to perform a physical action (close a relay, turn a motor, whatever) that takes some finite time to perform. If the receiving processor returns a command acknowledge to the requesting processor, and the requester takes this as a "command complete" indication, there may be timing problems. One way around this particular problem is to allow the command acknowledge to function as a "command complete," but delay the acknowledge until the function has actually been performed. Another solution is to have separate command acknowledge and command complete/command fail responses returned to the requester. Figure 5.5 illustrates these conditions.

One timing issue that seems to create a number of problems is the use of relays and solenoids. (A relay is a magnetic coil that, when activated, pulls in contacts to make or break a switch closure. A solenoid is a magnetic coil that pulls or pushes a cylinder to move some mechanism.)

The problem with relays and solenoids is that they take several milliseconds (several tens of milliseconds for a large one) to perform the action. The software must allow for this when activating a relay, or solenoid, and not expect the physical action to be immediate.

The real-world example of this involved two relays, one to start and one to brake an AC motor. There was hardware to prevent the brake relay from activating before the drive relay had opened. But there was no protection the other way, so a quick operator could stop the machine and restart it before the (brief) braking cycle was complete, destroying the drive relay. The solution was to add a software timeout that prevented a restart until after the braking cycle was complete.

Race Conditions

This special case of timing problems usually involves two things trying to happen at the same time. The hardware version of a race condition involves times usually measured in nanoseconds. Times for a software race condition are much longer, and are measured in instruction cycles, which are usually milliseconds, microseconds, or hundreds of nanoseconds.

Race conditions caused by interrupts are covered in the next chapter, so we'll look only at noninterrupt-based race conditions here.

Figure 5.6 shows a system block diagram and software flowchart for a simple embedded system. This system is a translator that converts serial RS-232 data from a host (such as your PC) to some other, undefined, format (such as the parallel interface of your printer). The embedded system receives and buffers data from the host, and transmits it to the peripheral at whatever rate the peripheral can take it. To prevent overflow, standard XON and XOFF protocol is used to control data flow. When the embedded system sends XOFF to the host, the host will stop transmitting data, although anything already buffered for transmission will be sent. When the embedded system sends XON, the host will resume transmitting.

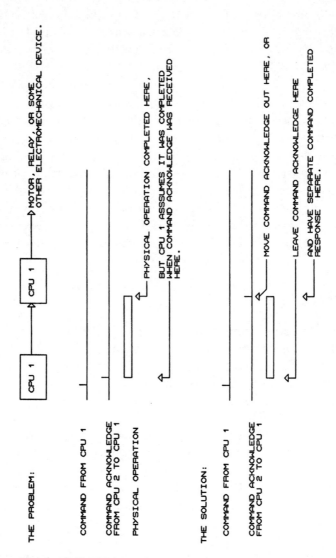

Figure 5.5 CPU-to-CPU Communication Timing

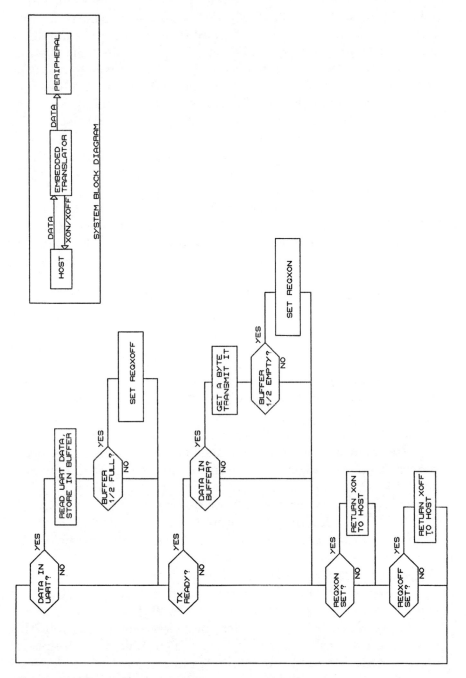

Figure 5.6 Simple Data Translator

The software is fairly simple and does not use interrupts. As each byte is received, it is placed in a buffer. If adding a byte to the buffer makes it half full, a flag called REQXOFF is set, to request that XOFF be returned to the host. The half-full point is used to request the XOFF since some additional data may be transmitted after the XOFF is sent.

After the receive data is processed (or if there is nothing in the UART), the peripheral interface is checked. If it is ready and if there is something in the buffer, a byte is sent to the peripheral. If the buffer emptied past the half-full point, an REQXON is set to request that the XON signal be sent to the host.

After the peripheral is serviced, the code checks for REQXON and sends XON if it is set, then checks for REQXOFF and sends XOFF if it is set. Then the entire loop repeats, forever.

The way this is intended to work is that the buffer will fill to the halfway point (assuming the peripheral is slower than the host), XOFF will stop transmission after a few more bytes, then when the buffer has been emptied back to the halfway point, XON will be sent to resume transmission.

What is almost guaranteed to happen sooner or later is this: the buffer will half fill and REQXOFF will be set. In the same pass of the loop, the peripheral will go ready and a byte will be transmitted to it. This empties the buffer to the halfway point, and REQXON is set. This is the race condition.

When we get to the bottom of the loop, the code checks for REQXON, finds it set, and sends the XON to the host (which already has transmission enabled). It then checks REQXOFF, finds it set, and sends XOFF to the host, which stops transmission.

Now the system is locked up. Transmission from the host has been stopped, but the translator thinks transmission is enabled (because the buffer is less than half full), so it never sends the XON request.

This problem occurs because the design assumes that REQXON and REQXOFF will never be set in the same pass through the loop. Although the code that sets REQXON and REQXOFF is many instructions apart, the two conditions get set at the same time as far as the on/off checking routine at the end of the loop is concerned.

There are a number of ways to fix this hypothetical problem. One would be to have XOFF sent when the buffer fills up to three-quarters full and XON sent when the buffer empties to one-quarter full. This provides a hysteresis that solves the problem unless the peripheral interface code can send multiple bytes in one pass of the loop, or unless the host response path can be busy for some time.

A more sophisticated solution would be to have a flag, set/cleared by the REQXON/REQXOFF checker, that keeps track of the XON/

XOFF state of the host. REQXOFF would not be set unless the buffer was at least half full and the host was on. Similarly, REQXON would not be set unless the buffer was at least half empty and the host was off.

This example serves to illustrate an important concept, which is *granularity*. For the purposes of this illustration, granularity is the window within which two events can happen and be considered simultaneous by the software.

We usually think of granularity as being a fixed time value, such as a regular timer tick, or a specific, fixed sequence of instructions. But in this case, the granularity is the time it takes to execute the software loop, and that time varies with the branches taken. For our data translator example, any time that both of the flags are set in one pass of the loop, *it appears simultaneous to the software*. While this may seem to be a "so what?" issue, it can have an impact on the repeatability of errors that you are trying to debug.

Take another look at the motor/shaft/mechanical assembly system. Let's say that our system is now a subsystem controlled by a host via an RS-232 serial port. The host system sends commands such as "Move out 3 inches" or "move home," which returns the mechanical assembly to the motor end of the shaft.

Our embedded processor saves commands received during a move and executes them after the move is complete. The checking sequence executed after a move looks like this:

If a "move x inches" command was received, execute the move.

If a "home" command was received, move home.

Now suppose that our subsystem is executing a move from 4 inches to 5 inches. During this move, the host sends a command to move home and then a command to move out 3 inches. The host wants the final position to be 3 inches from the end of the shaft, but what actually happens is that the final position is at home. The sequence looks like this:

Start move from 4 inches to 5 inches.

Home command received and stored for later execution.

Command to move out 3 inches received and stored for later execution.

First move (from 4 inches to 5 inches) completed

Move out 3 inches command executed, making final position 8 inches (5 inches to 8 inches).

Home command executed, leaving final position at home.

Now, obviously, this problem could be fixed by buffering commands and executing them in the order received. But it illustrates an important point, which is that the granularity that causes the race condition is the time required to complete a move, and this time varies over a wide range. This problem will show up only if the host sends a "home" followed by a "move x inches," and both are received during a move operation. If commands are sent infrequently or if the RS-232 interface is fairly slow, this problem may show up only during very long moves.

Synchronization Errors

The data translator example we looked at resulted in a synchronization error between two different systems; the host was waiting for the translator to send XON, and the translator thought it already had.

For an example of a synchronization error within a single system, look at Figure 5.7. This shows the state diagram of a simple burglar alarm. The alarm consists of two separate state machines. The simpler machine, the input monitor, looks at a keypad where the user can enter a code to arm or disarm the alarm. This state machine also monitors the state of the user's front door to see whether it is open or closed. When the user enters an arming code, the code sets a variable called ICODE to 01. When the door closes, ICODE is set to 02. When the door opens, ICODE is set to 03. Once ICODE is set to any of these values, the code waits for ICODE to go back to zero before changing it.

The main loop uses a variable called MODE. The values of mode are as follows:

MODE = 1, idle with the door open

MODE = 2, idle with the door closed

MODE = 3, user entered arming code, waiting for the door to open

MODE = 4, waiting for the door to close so we can arm alarm

MODE = 5, armed

MODE = 6, triggered (door opened while armed), wait for timeout or arm code

MODE = 7, 30-second timeout, alarm on

The way the alarm works is that it will not allow arming while the door is open. Once the door is closed, if the operator enters the arming code, the system goes to MODE 3, where it waits for the operator to exit the house. This is signified by the door opening and closing. Once the door has opened and closed, the alarm goes to state 5, where it waits for the door to be opened. This can be either a break-in or the user returning home, so the code waits 30 seconds. If the arming code is entered during that time, the system returns to idle state; otherwise the alarm is turned on.

Since ICODE is set on transitions (door opening, closing, or entry of arming code), the main state machine resets ICODE every time it sees ICODE indicate a change (by being 1, 2, or 3).

Now, let's look at the following sequence: the door is closed; the user enters the arming code, and walks to the door. Just as the user pulls the door open, the telephone rings. Closing the door, the user goes back to answer the call. Now the main state machine is in state 5 (ARMED), and ICODE is 0. Finishing the telephone conversation, the user realizes that the alarm is armed, so she enters the arming code to disarm it. The input monitor detects the arming code and sets ICODE to 1. But the main state machine was written on the assumption that the house must be empty in ARMED mode, so it doesn't have an exit path from ARMED except when the door opens.

When the user enters the arming code and nothing happens, she tries opening the door, thinking this will trigger the alarm and allow her to disarm it. What actually happens is that the input monitor is stuck with ICODE = 1 (from the arming code entry) and will ignore the door open until the main code resets ICODE to 0. However, the main state machine doesn't recognize ICODE = 1 as a valid code while in the ARMED state, so it will never reset ICODE. The result is a permanent lockup.

There are a number of ways to fix this particular problem, but the point is that two state machines have ended up in an illegal combination that the code does not account for. All code has states, explicitly or implicitly, based on the values of various variables. The more variables there are, the more possible combinations exist. Be sure that they are all accounted for, or else be sure that the code cannot get into an illegal state.

Missed Deadlines

An old adage in the embedded world is that "The right answer, late, is the wrong answer." Deadlines can be broadly divided into two categories: *hard* and *soft*. A hard deadline is a deadline that must be met, every time, or the system will fail in some way. There is a lot of debate about what constitutes a soft deadline, but we'll define a soft deadline

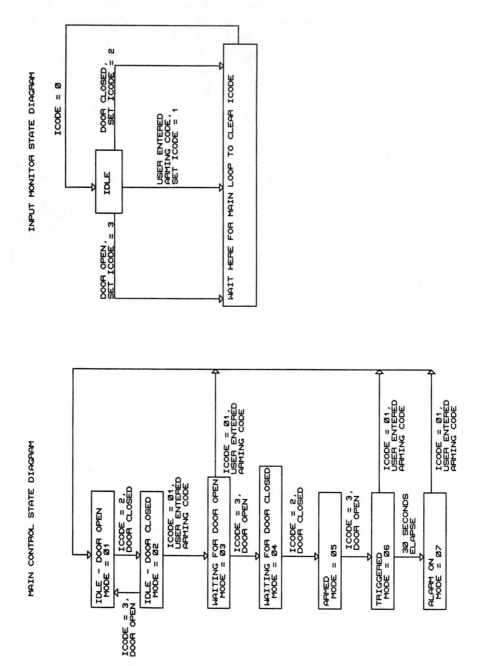

Figure 5.7 State Diagram for Burglar Alarm Example

as a deadline that can be missed occasionally, as long as the processor, on average, keeps up.

A simple example is a UART, like that used in the programmer. At 9600 baud (assuming one stop bit and no parity), a byte takes just over a millisecond to transmit. The programmer software that handles data from the UART just puts it into a buffer, so it can be processed later. As it happens, this code is in an interrupt service routine, but that is immaterial to this discussion.

The UART receives a byte and puts it in a register. The byte will be held in the register until another byte is assembled, and then the new byte overwrites the old one. If the incoming data is continuous, this will take one byte time, or about a millisecond. This, then, is the deadline for reading a byte once it is placed in the hardware register. Figure 5.8 illustrates this timing.

This is an example of a hard deadline. If the processor doesn't get to a new byte in a millisecond, the data will be overwritten by the next byte, and the old one will be lost.

The process of taking data out of the receive data buffer is an example of a soft deadline. The buffer is 256 bytes long, so if the code that processes received bytes (takes them out of the buffer and does something with them) is busy when a new byte comes in, it is not a problem. Additional bytes are received and fill new locations in the buffer. When the processing code gets to them, they will still be there. As long as the average processing time is less than a millisecond, everything will be fine.

A soft deadline can, under some circumstances, turn into a hard deadline. If the processing code for the programmer stays busy for too long, the buffer will fill up. When this happens, the soft deadline becomes a hard deadline. If the processing code doesn't start taking data from the buffer within 1 millisecond (the byte transmission time again), the next byte that is received will overwrite data in the buffer. In some systems, the new byte would not be stored if the buffer was full. In either case, a byte is lost.

Tracing Data Flow

The normal debugging method involves tracing execution of the program to find where it goes wrong. In some cases, it is instructive to trace the flow of data instead.

Figure 5.9 shows two flow diagrams for a segment of the programmer code. The first diagram, Figure 5.9A, traces program flow for the numeric entry that selects a target device type (i.e., 2764, 27256,

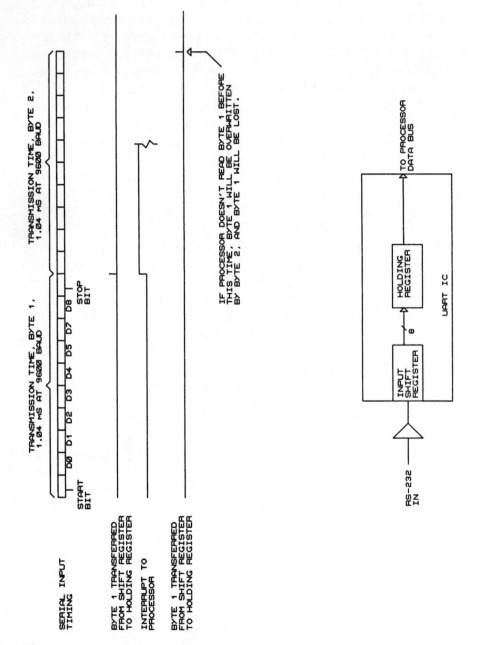

Figure 5.8 UART Receive Timing

etc.). At the point where this diagram begins, the programmer has already received a #T command and is waiting for an ASCII number that represents the device type to be selected.

As shown in the figure, the background code detects data in the receive FIFO (placed there by the UART ISR) and calls the RXPROCESS routine. RXPROCESS checks flags to determine what is expected and finds that the code is in a state where the #T command has been received, that a delimiter between the command and the data has been received, and that data is expected. RXPROCESS then passes control to RXDATP, which is the routine that handles data for commands that require it. RXDATP verifies that the data is not another delimiter, that it is a valid ASCII number, and then calls ASCHEX to convert the character to a hex value. This number is then placed into the variable (DEVTYPE) that holds the device type.

The second diagram in the figure, Figure 5.9B, shows the data flow. The value in the receive FIFO is moved to a variable, CHAR by RXPROCESS. The contents of CHAR is moved to ASC by RXDATP for conversion to hex. ASCHEX converts ASC to hex, leaving the result in HEX, and RXDATP takes the contents of HEX and moves it to DEVTYPE.

In most cases, tracing program execution is the best way to find a problem. But there are times when following the data flow makes a solution more obvious.

Following are some tips you can use to make software debugging easier:

1. Add debug tools. The #XB and #XM commands, and the trace buffer, are in the programmer code permanently. If a problem shows up at a later date, the tools to work on it are already there. Some programmers leave a ROM monitor in their code and let it get control with a special command or when a particular shunt jumper is installed on powerup. Just make sure it doesn't activate unexpectedly—such as when a customer makes a typing error.

2. As discussed in Chapter 1, most processors have a single-byte (or single-word, for wider buses) opcode that performs a software interrupt. If you fill all the unused ROM space with this opcode, then berserk code will probably hit this instruction eventually, and you'll be able to tell what happened. But make sure you actually handle the interrupt when it occurs.

3. Be sure your functional test is complete.

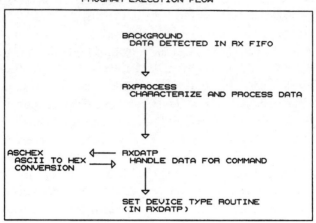

(a)

(b)

Figure 5.9 Tracing Program Flow Versus Data Flow

4. Any time you have one state machine dependent on the state of another state machine, be sure there are no illegal combinations, or be sure you handle them, or be very sure the code cannot get into them.

In Chapter 6, we'll look at interrupts, which is what makes most real-time systems real time.

6

Debugging the Interrupts

"Got a minute?" Cody asked as he dropped into the chair beside Matt's desk.

"Sure. What's up?"

"I showed the programmer to Josh and one of the marketing guys, and they want some changes that I'm not sure how to accomplish. They say that the messages are too slow during programming, it makes the system look sluggish. And they want to be able to send any command from a script file without worrying about the programmer falling behind and losing data."

"Let's take them one at a time," Matt said. "How are your messages transmitted?"

Cody pulled several stapled sheets from a folder. "Here's the pseudocode for the transmit portion of the background loop," he said, pointing.

> If UART Tx register empty (ready to transmit),
>> If there is data in the Tx FIFO,
>> Transmit the next byte from the FIFO.

"So you do your transmitting from the background loop," Matt said.

"Sure. I thought speedy transmitting was less important than getting the PROMs programmed."

"But the way your code is structured, you can't transmit more than one byte for every pass through the background loop. You also program one byte for each pass through the loop, and the loop doesn't execute while a byte is programming. So if you are programming a chip that needs, on average, 5 milliseconds per byte, your maximum transmission rate is, um, 200 characters per second. No wonder it looks sluggish."

"So I have to service the UART transmitter from the ISR just like the receiver, right?"

"That sounds like the best solution."

"But my ISR is already too complicated. That's why I have problems with executing back-to-back commands."

"Do you execute your commands in the ISR?"

"Sure. And if it's a fill buffer command that takes a while to execute, I miss the next couple of characters because they come in while I'm doing the fill."

"I'd move the command processing to the background and make the ISR as simple as possible."

Matt picked up a marker and wrote on the whiteboard:

New ISR

Read received byte from UART.

Store in Rx FIFO.

Update Rx FIFO write pointer.

Send EOI to interrupt controller

Return.

"How do I add the transmit stuff?" Cody asked.

Matt erased what he had written and started over:

New ISR

If UART Rx ready bit set (Rx data available),

    Read received byte from UART.

    Store in Rx FIFO.

    Update Rx FIFO write pointer.

    Send EOI to interrupt controller, then return.

Otherwise (there was no Rx data),

    Write a byte from the Tx FIFO to the UART.

    Update the Tx FIFO read pointer.

If the Tx FIFO read and write pointers are equal (FIFO empty),

    Clear UART Tx interrupt bit.

    Send EOI to interrupt controller and return.

"Okay. I'll go work on this."

The interrupts in an embedded system are, in many systems, the key to doing things in real time. They also add a level of complexity to the system that complicates the debugging process. Although they are sometimes the only way to make sure that a particular device or task is executed in a predictable fashion, they tend to make the overall system less stable and less predictable.

Interrupt Overview

An interrupt signals an external event to the microprocessor. It may indicate that a particular amount of time has elapsed, or that a user has pressed a button, or that a mechanical arm has moved another step. In general, interrupts are used to be sure that something gets serviced right away.

Let's look at the programmer hardware as an example of the mechanics of servicing interrupts. When the UART receives a byte of data, it generates an interrupt on the 80188 INT0 pin. The 80188 internal interrupt controller sees the interrupt pin go active and interrupts the processor. The processor finishes executing the current instruction and then executes an interrupt acknowledge cycle, which is recognized by the interrupt controller. During the acknowledge cycle, the interrupt controller provides an interrupt vector to the 80188. The interrupt vector is a byte, and, in the case of the 80188 INT0, has a value of 12 (0C hex). The 80188 multiplies this by 4 to create a vector address, then gets an interrupt vector from the new address (00030 hex for INT0). The first 1k of memory, 00000 through 003FF, is the interrupt vector table for the 80188 and contains interrupt vectors for the various interrupts. In the case of INT0, locations 00030 through 00033 contain a pointer to the UART ISR in the EPROM. The processor saves the return address (the next instruction that would have executed if the interrupt hadn't occurred) and branches to the address it reads from the vector table. Figures 6.1 and 6.2 show the 80188 interrupt sequence.

The ISR saves the CPU *context* and any registers that it will use, does whatever it needs to do to handle the interrupt, and then informs the interrupt controller that the interrupt has been serviced. This is necessary because the interrupt controller remembers which interrupt is in process and will not permit another one to be serviced until it has been told that the previous interrupt was serviced. The CPU context is defined as the current state of the CPU (flags, condition codes, etc.).

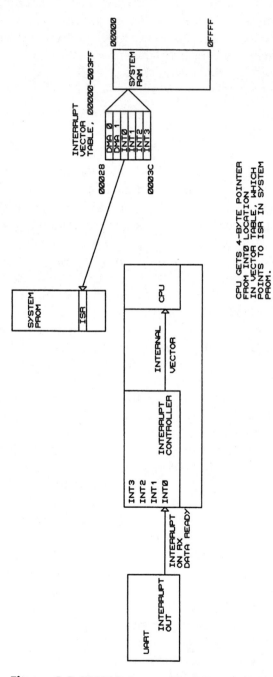

Figure 6.1 80188 Interrupt Block Diagram

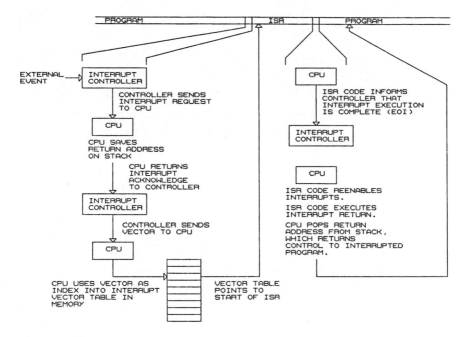

Figure 6.2 80188 Interrupt Sequence

The last thing the ISR must do is restore the state of the CPU to whatever it was before the interrupt occurred. This means restoring the saved registers and reenabling interrupts. The CPU disables interrupts during interrupt servicing, so the ISR must tell the CPU when to turn interrupts back on. This is the End of Interrupt (EOI).

If everything was done right, the CPU will resume execution at the place it left off when the interrupt occurred, as shown in Figure 6.2, and that section of code will not even know the interrupt happened.

It is important to note the difference between the interrupt input and the interrupt to the CPU. To illustrate this, Figure 6.3 shows an 8088 processor using an 8259 interrupt controller. The 8259 accepts up to eight external interrupt inputs. But the 8259 provides only one interrupt input to the CPU. When a device requests an interrupt, the 8259 generates an interrupt to the CPU and then gives the CPU a vector when the CPU performs an interrupt acknowledge cycle. The vector is based on *which* interrupt input is active.

The peripheral that generated the interrupt does not know when the CPU is interrupted or when the interrupt acknowledge occurs. The

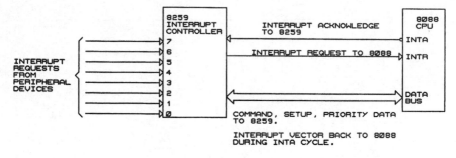

Figure 6.3 8088 with 8259 Interrupt Controller

first time the peripheral knows that its interrupt has been handled is when the ISR services the hardware. In fact, if the interrupts are prioritized, an interrupt from a low-priority peripheral may not be serviced for a while, if a higher priority interrupt is in progress.

The 80188, just described, has several interrupt inputs, both from external and internal sources. The CPU core, however, still gets a single interrupt and a vector from an internal controller.

Since the interrupt vector table is usually in RAM memory, the CPU must write the vector table before the first interrupt occurs. In the case of the programmer software, a copy of the interrupt vector table is stored in PROM and is copied to RAM after powerup reset before interrupts are enabled for the first time.

Some processors, such as the 8031, do not use a vector table. Instead, an interrupt sends the processor to actually execute code at a particular address, as shown in the following table:

8031 Interrupt Addresses

Interrupt Source	Vector Address (hex)
I0	0003
Timer 0	000B
I1	0013
Timer 1	001B
UART/USART	0023

These memory locations would typically contain a branch instruction that jumps to the actual ISR somewhere else in ROM, although a very simple ISR might fit in the 8-byte interrupt address.

Some processors, such as the Z-80 family, allow an instruction to be presented to the processor during the interrupt acknowledge cycle. This will typically be a 1-byte software interrupt instruction that has the same effect as an interrupt vector.

Interrupts come in two flavors: edge and level sensitive. *Edge sensitive interrupts* are activated when a signal changes state. An edge sensitive interrupt typically must change from the inactive state to the active state, then be held in the active state until serviced. A *level sensitive interrupt* will be serviced as long as it is in the active state. Some processors, such as the 80188 and 8051, have external interrupts that can be programmed to be either level or edge sensitive.

Interrupts on some processors can be prioritized. When prioritizing is used, a high-priority interrupt can interrupt the ISR for a lower-priority interrupt, but not vice versa. Methods for prioritizing interrupts, and the flexibility those methods provide, vary from processor to processor.

Interrupts are asynchronous: they can occur at any point in code execution. Even a regular interrupt, such as a timer tick, is asynchronous with respect to the rest of the code. This one fact is probably, directly or indirectly, responsible for most of the difficult problems in interrupt-driven systems.

Debugging Interrupts

The traditional tools for debugging software are less useful for debugging interrupt problems. The most common tool for users of emulators or ROM monitors is the breakpoint. When a breakpoint is executed, the processor stops or transfers control to the monitor program. The problem is that interrupts keep coming, and they stack up. As soon as you try to step the code or resume execution, all those stacked interrupts want attention at once. All the hardware events they represent occurred several seconds or minutes ago. In addition, because interrupts are asynchronous, the actual cause of a problem may have occurred long before the symptoms you can trigger on.

Potential Interrupt Problems

The interrupt system is susceptible to all the hardware and software problems to which the rest of the system is susceptible, and also to some unique faults that occur nowhere else.

Stack Control

When the ISR is entered, it is usually necessary to save all the registers that will be used and restore them on exit. This is usually performed by pushing the registers onto the stack and popping them off before returning. Of course, the return address is also placed on the stack when the ISR is executed, so if there are more pushes than pops (or vice versa), the ISR will return to some unknown location. This condition is usually obvious—the processor just goes off "into the weeds" as soon as the interrupt is enabled. A less obvious condition is where the registers are popped in a different order than they were pushed. The following 80188 code fragment illustrates this:

```
; example ISR
push ax ; save registers on
push cx ; ISR entry.
push dx
.
. ; ISR processing code
.
pop cx ; restore registers
pop dx ; prior to exit
pop ax
```

In this case, the registers were pushed in the order AX, CX, DX, but popped in the order CX, DX, AX. This has the effect of swapping the contents of CX and DX when the ISR returns. The symptom this produces can be very subtle. Let's say that the background code doesn't use DX or CX, but some seldom-used subroutine does. In this case, there may not be any problem until the ISR occurs during that subroutine. Or, suppose the same background code is modified so that it uses CX or DX. The new code will appear to have a bug, when it is actually the ISR that is causing the error.

Some designers get around this problem by using a standard save/restore macro that always saves all the registers. The only problem with this solution is that you may end up saving registers you don't use in the ISR, and that takes time.

This problem is unlikely if the ISR is written in an HLL; the compiler takes care of saving and restoring the registers.

Shared Resources

This can be the cause of symptoms that are serious but difficult to isolate. This class of problems can be further divided into shared variables, shared I/O, and shared subroutines.

Shared Variables

Take a look at the two pseudocode fragments below:

Non-ISR code:

```
1 Read variable X
2 Decrement
3 Store result back at variable X
```

ISR code:

```
Read variable X
Increment
Store result back at variable X
```

Let's say that X is counting the number of bytes in some hypothetical buffer. The ISR code puts a byte in the buffer and increments the count to indicate the change. The non-ISR reads a byte from the buffer and decrements the count.

Say that X starts out with a value of 4. The ISR code puts a byte in the buffer and increments X to 5. The non-ISR code then reads a byte and decrements the count back to 4. But if the interrupt occurs while statement 2 in the non-ISR code is executing, the value of X will be corrupted. First, the non-ISR code reads X, which is 4. Then the ISR occurs and increments X to 5. After the ISR completes, the non-ISR code finishes, and stores 3 in the buffer.

A similar problem can occur with binary variables (semaphores) used to control access to some common resource, such as a printer. Say that the variable FLAG indicates availability of the printer. If FLAG is cleared, the printer is available and any subroutine in the code can use it. The subroutine that wants the printer must set FLAG to tell all other subroutines that the printer is in use, and resets FLAG when finished with the printer so that other subroutines can have it. Now look at the following pseudocode:

Non-ISR code

```
1 Read FLAG
2 If cleared, set FLAG to indicate printer is in use
```

ISR code

```
Read FLAG
If cleared, set FLAG to indicate printer is in use
```

Both the ISR and non-ISR code use the same logic to get control of the printer. If the interrupt occurs between statements 1 and 2 in the non-ISR code, both the ISR and the non-ISR code will think they have control of the printer, and a conflict will result.

This problem has several solutions. Some processors have a test-and-set instruction that allows semaphores (such as FLAG in the second example) to be safely modified. These instructions will typically read the memory location, test/set the appropriate bit, then write the new value to the memory location. The instructions are typically locked or indivisible, meaning that the instruction, once it begins execution, will complete regardless of interrupts, DMA requests, and other external events.

A second solution to this problem is to disable interrupts around the read/test/write sequence in the non-ISR code, as illustrated below:

Protected Non-ISR code

```
Disable interrupts
1 Read FLAG
2 If cleared, set FLAG to indicate printer is in use
Reenable interrupts
```

The simplest solution, and recommended whenever possible, is simply to avoid shared variables. To illustrate how this might work in the first example above (the buffer problem), let's replace variable X with two pointers, a read pointer and a write pointer. Both pointers start out pointing to the beginning of the buffer. As the ISR puts bytes in the buffer, it increments the write pointer. To find out whether there is something in the buffer, the non-ISR code tests the pointers to see whether they are equal; if they are, the buffer is empty. If the pointers are unequal, the non-ISR code reads a byte and increments the read pointer.

Although this is a simplified explanation and leaves out some things (what do you do when one of the pointers gets to the end of the buffer?), it serves to illustrate the concept. Note that the ISR and non-ISR code never write the same pointer. This prevents the conflicts that occur with the shared variable. The PROM programmer code uses this method to handle UART receive and transmit data.

Even if the non-ISR code never writes a variable that is written by the ISR code, it is possible to have a sharing conflict. Look at the following pseudocode:

Non-ISR code

```
Read variable X
If X = 1, do something
```

```
Read variable X
If X = 2, do something else
Read variable X
If X = 3, do a third thing,
etc.
```

If the variable X were controlled by the non-ISR code, this code would do only one thing each time it executes, based on the value of X. However, if the variable X is set in an ISR, and if the interrupt occurs while this chunk of non-ISR code is executing, bizarre and apparently impossible results may be seen, since two of the "somethings" may be performed in one pass of the code. Since X is read every time it is tested, it can be changed by the ISR between the beginning and the end of the routine.

This problem is not as unlikely as it may appear. The 8031 micro-controller, for example, does not have a compare instruction. If X were a byte variable to be compared with three constants, the code would have to load X into the accumulator register and perform a subtract or exclusive-OR operation. Either of these alters the contents of the accumulator, so it must be reloaded before being tested again. Also, some compilers can create this sort of problem.

The fix is obvious; if you are reading a variable that is set by an ISR, and you need to use it more than once outside the ISR, read it once and store it in a temporary location, then use the copy.

Shared I/O

This problem is similar to the shared variable problem, but may be more difficult to solve. Figure 6.4 shows a control register used in the programmer example. This 8-bit register provides six control bits to the PROM programming connector and also controls the bicolor LED.

Let's say that the LED is controlled by an ISR while the control bits are managed by non-ISR code. Since this is a write-only register, a mask byte, which we'll call CMASK, must be stored in memory. If the ISR wants to set the LED green, it might use the following code:

ISR code to set LED green
```
AND CMASK with 0FCh to clear the LED bits.
OR 02 with CMASK to set the LED green.
Write CMASK to the control register.
```

Now, suppose the non-ISR code wants to set the CE output low. It might do this:

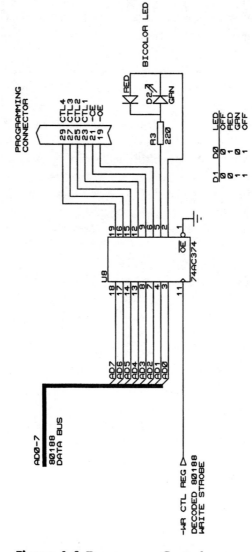

Figure 6.4 Programmer Control
Register

```
Read CMASK
AND with OFB to set CE low.
Write result to CMASK and to control register.
```

Now we have a problem similar to the shared variable problem. If the interrupt occurs between the non-ISR read of CMASK and writing the modified value to CMASK, the LED will flicker briefly green (probably too fast even to see the flicker), then revert to whatever it was before the ISR tried to change it.

There are a couple of fixes for this problem: the most obvious is to disable/enable interrupts around the non-ISR CMASK/LED register update. Another solution would be to define a second mask variable, which we'll call LEDMASK, for the LED only. Then our pseudocode looks like this:

ISR code
```
Set LEDMASK to 02 to make LED green.
```

Non-ISR code
```
Read CMASK.
AND with OFBh to set CE low.
Write result to CMASK.
OR with LEDMASK to update LED.
Write result to control register.
```

The drawbacks to this are that the LED color changes only when the non-ISR code changes one of the other control bits, and you always have to remember to include the value in LEDMASK when updating the control register.

A similar solution would be to create CONTROLMASK instead of LEDMASK and have the non-ISR code update this variable, but not update the control register. The ISR code would then keep the LED value in MASK, and update the LED register, including the value in CONTROLMASK. This, of course, reverses the problem in the previous solution: the non-ISR code cannot update the control bits immediately but must wait until the next ISR update of the LED register.

Some peripheral ICs, such as the Zilog Z8530 (serial communication controller) and Z8536 (timer/counter/parallel I/O), have more internal registers than they have external addresses. Registers in these devices are manipulated by first writing a value to an internal address register, and then reading or writing data to a different address to access the selected register. Figure 6.5 shows a simplified block diagram of how these devices are structured.

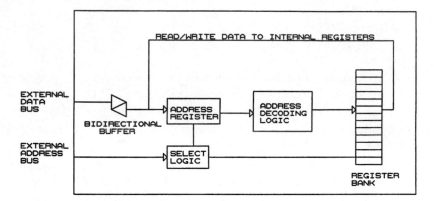

Figure 6.5 Peripheral IC, Such as the Z8530, with Internal Address Selection Register

Pseudocode to access a register in these devices looks something like this:

```
Write number of desired register to address register.
Read/write data address to access selected register.
```

A problem occurs if an interrupt happens between the above two operations, and if the ISR also must manipulate the peripheral device. The non-ISR code will set the address register to select whatever data register it needs to access. Then the ISR gets control and accesses some register. Then the ISR returns and the non-ISR code completes its access, but now the address register has changed, so the non-ISR code reads (or writes) the wrong register.

Devices that have this characteristic are typically high-integration parts with a number of functions, and there may be no way to avoid having both ISR and non-ISR code access the device. In this case, the only reliable solution is to bracket the non-ISR access with an interrupt disable/enable pair.

Shared Subroutines

The PROM programmer that Cody is building has a subroutine that converts hex data to ASCII for display. Data is passed to this routine through a global variable, HEX. The subroutine puts the resulting ASCII in a global variable, ASC. Now, suppose that this is used by both ISR and non-ISR code. Then suppose that it has been called by non-ISR code, and then the ISR that uses the subroutine is activated by its interrupt.

When this happens, the ISR code will write to HEX whatever data it wants converted, call the conversion subroutine, and then read the result from ASC. When it returns, the values of HEX and ASC have been corrupted (as far as the non-ISR code is concerned), and the wrong values will be returned. This is a software version of the address/data register problem just described.

Unlike the hardware registers, this problem can be solved without disabling interrupts. The straightforward solution is to make the conversion subroutine reentrant, passing the hex input data and the ASCII return data on the stack. Any local variables must also be dynamic, meaning they, too, are stored on the stack (or in memory that is allocated each time the subroutine is called). This way, when the subroutine is interrupted and called again by the ISR, the original input data and any half-completed output data will be saved on the stack.

Some microcontrollers, such as the 8031, have a limited stack that makes normal reentrancy impossible. One solution to this is to have a register that points to a data area where the input and output values reside, and where space is allocated for any temporary variables needed by the conversion subroutine. The ISR then saves this register on entry, calls the subroutine, and restores the register on exit. As long as the ISR and non-ISR code don't try to use overlapping data areas, the data will be preserved.

Interrupt Stackup

Interrupt stackup problems can take three forms: latency problems, stack overflow problems, and execution delays.

Latency is the time from when an interrupt occurs until it actually gets serviced. In general, it is best to assume that, at some point in the execution of your code, you will have all interrupts occur at the same time. The result of this is that the delay in returning to the non-ISR code will be total execution time of all the interrupts, as shown in Figure 6.6. Here we have a system with three interrupts. Interrupt 1 occurs first, and while interrupt 1 is executing, interrupt 2 occurs. As soon as the ISR for interrupt 1 is finished, the ISR for interrupt 2 executes. A similar sequence occurs for the third interrupt. The result is that the background code does not get control again until *all three* ISRs have completed. Note that the three interrupts did not have to occur at the same time to become stacked.

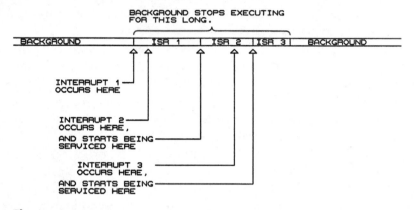

Figure 6.6 Interrupt Stackup

If you believe you have latency problems, you can determine the latency for any given interrupt, if it is level-sensitive. Level-sensitive interrupts are asserted by some event and then cleared by the ISR, usually by reading/writing hardware. Thus the interrupt is active from the time it is asserted until it is serviced. Connecting a DSO set for repetitive mode to the interrupt line and triggering the DSO when the interrupt signal goes to the active state will give you a good picture of the minimum and maximum latency. Each new interrupt will trigger the DSO, which will overlay the new pulse width on the existing display. Extremely long latencies will show up as a very long pulse width. The one caveat to this method is that the DSO will take some time to paint the screen and may miss interrupts that occur during the screen update.

The latency of servicing an interrupt is only half the issue: in most cases, the interrupt condition is cleared early in the ISR, but there may be considerable processing in the ISR after that point. It is often useful to know how long the ISR takes.

Measurement of the entire ISR service time can be accomplished by adding a hardware trigger just before the ISR exit. This can consist of setting, then clearing, a port bit on a microcontroller or on a PIO chip, setting/clearing a spare bit in a hardware register, or writing/ reading an unused location that will strobe the output of an I/O address decoder. To use this method, connect one channel of the DSO to the interrupt line, as before, and connect the second channel to the hardware trigger. Now when the interrupt occurs, the hardware trigger will indicate when the ISR is completed.

Some processors have internal peripherals that can generate interrupts. Examples of this are the timers and DMA channels on the 80188, and the timers on the 8031. Often these peripherals do not provide an interrupt signal outside the chip so that the latency can be measured. The execution time of the ISR, however, can still be measured. To do this, two hardware triggers are needed, one at the beginning of the ISR and one at the end. On a microcontroller, PIO chip, or hardware register, you can set a bit at the beginning of the ISR and clear it at the end. The DSO, again in repetitive mode, is used to capture the width of the pulse, which corresponds approximately to the execution time of the ISR.

Limited Stack

Some processors have a limited, fixed stack. The PIC17C42, for example, has a stack limited to 16 levels. If the total of your subroutine calls and ISRs exceeds this, the stack will overflow. Since the return addresses on the stack all point to valid code somewhere, the code will not necessarily go berserk.

The 8031 family of processors has stack limitations of a different type. The stack on an 8031 is in memory, so the stack is limited only by memory size. However, the internal memory on an 8031 is very limited, so it is easy to have the code grow down into the variable area and overwrite data.

If you suspect problems with stack size limitations, you might want to check the stack pointer in each ISR when it is called (on processors where you can get to the stack pointer). If you find the stack pointer at the end of the stack, set a port bit or send an action code, or do something to let the rest of the world know about the problem. A better solution is to assume that an interrupt will occur while you are at the deepest level of subroutine calls, and calculate what the result will be on the stack pointer. Fortunately, processors with extreme stack limitations are typically used in less complex applications, where it is easier to predict what the stack will do.

UART Transmit Lockup

Let's look at the code for the UART transmitter in the programmer:

The UART itself has a bit in a hardware register, TIENAB, that allows the UART interrupt output signal to go active when the transmit register is empty.

The UART transmit code is divided into two portions: a subroutine called SNDMSG, and the UART ISR code. SNDMSG is called with a pointer to the message to be sent. Messages are terminated with a byte of 0FFh. A simplified description of SNDMSG might look like this (TWPOINT is the Tx FIFO write pointer):

```
Get byte at M[essage Pointer].
If byte < FFh (message terminator),
Store byte in Tx FIFO[TWPOINT].
Increment TWPOINT, wrapping to 0 if overflow past 0FFF
  (1k).
Otherwise (byte at Message Pointer = FFh, indicating end
  of message),
Set TIENAB.
Return.
```

The transmit portion of the UART ISR looks like this (TRPOINT is the Tx FIFO read pointer):

```
If Tx interrupt active (meaning transmit register is
  empty),
If there is data in the Tx FIFO,
Transmit byte at Tx FIFO[TRPOINT].
Increment TRPOINT, wrapping to 0 if overflow past 0FFF
  (1k).
If FIFO now empty, clear TIENAB.
```

Now we'll analyze this briefly. The 16550-type UART used in the programmer is typical of UART ICs. It has a single interrupt output pin that is shared between receive and transmit. The software can enable and disable interrupts for either receive or transmit. If both interrupts are enabled, the interrupt pin will be active for either receive or transmit ready, and the software must read a register in the UART to determine which condition is causing the interrupt.

The receive interrupt function is straightforward—the interrupt is active when a new byte is in the receive holding register, and it stays active until the software reads the byte. The receive interrupt can typically be enabled all the time, since it generates an interrupt only when data is actually available.

The transmit interrupt is a different story. The transmit interrupt indicates when the UART is ready to accept data, whether there is any data to be sent or not. The *default* state of the transmitter, when it is idle, is to generate an interrupt. But the software needs to get an interrupt from the transmitter only when the UART is ready for data *and* when there is data to send. Otherwise, when there is no data to send, the interrupt signal will be continuously active and the software will spend all its time servicing the interrupt.

The way around this is to enable the transmit interrupt when there is data to send and disable it when all the data has been sent. But this requirement causes a potential problem. The non-ISR code must set the TIENAB bit and the ISR must clear it when a message is finished transmitting. So a potential race condition can exist. In the programmer, this is handled two ways:

1. The transmit buffer is a FIFO. The ISR code checks for data in the FIFO by checking for inequality between the read and write pointers. Since the write pointer for new data is set up well before the instruction that sets TIENAB, the ISR won't find the buffer going empty just as a new message is placed there.
2. The non-ISR code sets TIENAB when done loading a message, without checking it first. This prevents a race condition that could occur if the transmit interrupt occurs between the test and the set operations. TIENAB is a bit in a UART hardware register, so writing a one when it is already a one will not cause problems.

Potential problems with any hardware, such as a UART, that needs control of an interrupt enable bit include:

- A race condition between test and set of the enable bit in the non-ISR code
- The receive interrupt enable and possible other enable bits (such as modem control) may be shared in the same UART register as the transmit enable bit. If the other bits are sometimes turned off, the code cannot set the transmit bit without reading the register (or a mask of it) and ANDing or ORing the new

value in. The time between read and write has the potential for a race condition.

- Even with the arrangement used in the programmer, there is the potential for a problem if the message to be sent is very short (say, 1 byte) and the transmit ISR occurs before SNDMSG sets TIENAB. The ISR would see 1 byte in the FIFO, clear TIENAB, and return. SNDMSG would then set TIENAB even though the FIFO is now empty. To prevent this problem, the transmit ISR clears TIENAB if it finds an empty FIFO.

Interrupt Time

Most of the potential interrupt problems we've looked at have involved resources that are shared between the ISR and non-ISR code. It is possible for an interrupt to create problems completely unrelated to anything the ISR does or any resource it uses (except one).

Figure 6.7 shows how the asynchronous nature of interrupts can affect code that is unrelated to the interrupt. Here we have a timer IC, such as the 8253, connected to a processor. The gate input to the timer is controlled by 1 bit of a PIO IC or a register. When the gate is high, the timer increments, and when it is low, the timer holds the current count. The 16-bit timer is counting some external event, such as motor encoder pulses, regular ticks from another timer, or maybe items moving down a conveyor belt. Our hypothetical system also has an interrupt that is completely unrelated to the timer operation.

As shown in the figure, the non-ISR code, for whatever reason, wants to read the timer contents. The problem with timers is that they may increment while reading, and the processor may end up with a mix of the old and new bits. So our non-ISR code uses the gate to stop the timer while reading the count.

As can be seen in the timing diagram, if the interrupt occurs during this read, and the ISR does not return until after the next external event occurs, a count is missed.

To fix this, we try leaving the gate active all the time and just reading the high and low bytes separately. Now a different problem is introduced. The processor reads the low byte of the count (32FF), and gets FF. Then the interrupt occurs, and after the ISR, the processor reads the high byte of the count (now 3300) and gets 33. The assembled word is now 33FF, completely incorrect.

The fix for this is to use a timer that will allow you to freeze the count in a separate register, or to read the low/high pair twice and

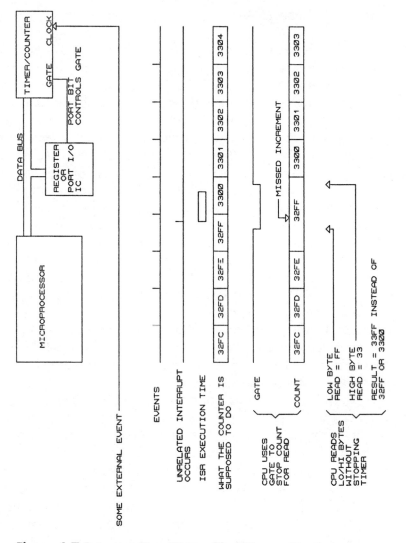

Figure 6.7 Interrupt Time Usage. The ISR execution time causes problems with a completely unrelated function.

compare them. More important, though, is this point: any time you have an interrupt-driven system with operations that are divisible but should not be, you have the potential for problems, because there is one global resource that *all* interrupts must use—time.

Prioritizing Interrupts

Nearly all systems have interrupts of different priorities. The 80188, for example, has several internal interrupt sources with differing priorities. It is not uncommon for a software design to leave interrupts disabled while in an ISR. In this scheme, a high-priority interrupt that is asserted at the same time as a low-priority interrupt will be serviced first. But a high-priority interrupt that occurs after a low-priority interrupt is acknowledged will be held off until the low-priority ISR is completed. This has the effect of making the low-priority ISR a higher priority.

In some systems, this is unacceptable. For example, you might have a processor that gets a continuous data stream from an RS-232 interface and also sends data to a printer. You would probably set the RS-232 receive interrupt to be a higher priority than the printer interrupt because you cannot afford to miss any serial data, but it's okay for the printer to run slow sometimes. If the printer ISR for some reason takes longer than the time needed to receive a serial byte, then the printer ISR can cause missed data.

The solution is to add an additional level of prioritization called *interrupt nesting*. In interrupt nesting, an interrupt can itself interrupt the ISR of another interrupt of a lower priority. In our example, the RS-232 ISR would be higher priority than the printer ISR.

To implement interrupt nesting, you need an interrupt controller that can support it. Different processors and interrupt controllers support nesting to various degrees. The 8018x processors do support this feature with the internal controller. The commonly used 8259 interrupt controller also supports nesting. The 8031 processor supports two levels of priority for nesting. The 68000 family parts (which we'll talk about later) has a prioritizing scheme that makes nesting nearly automatic.

The disadvantage to nested interrupts is that the ISRs themselves become susceptible to the same problems with divisibility, shared resources, and time that the non-ISR code has. In addition, the potential for stack overflow increases with nested interrupts. If you use nested interrupts, it is more important than ever to minimize the ISRs, thereby minimizing ISR execution time.

Problems with Vectored Interrupts

Of course, all interrupt systems are susceptible to the problems already described. Here we'll take a look at problems that are less general.

A system using several individual, prioritized interrupts (like that of Figure 6.3) can hang up with one interrupt never being serviced. If you find an interrupt that is not being serviced, first look to see if that particular interrupt line is always high. If not, the peripheral, whatever it is, is not requesting an interrupt. If the interrupt line is always high, suspect that the interrupt is masked or that there is a priority problem. Note that when interrupts are nested, if the interrupt controller thinks an intermediate interrupt is still being serviced, all lower priority interrupts are blocked, but all higher-priority interrupts *will continue to be serviced*.

Since each interrupt in an 80188 or 8259-type system has a priority, the ISR must tell the interrupt controller when it is finished servicing the interrupt. If the ISR fails to send the EOI to the controller, or gives the controller the wrong EOI code, a lower-priority interrupt may never be passed through, since the controller believes it is still servicing a higher-priority interrupt.

Daisy-Chained Interrupts

Figure 6.8 shows a block diagram of a microprocessor using daisy-chained interrupts. In a daisy-chained system, each peripheral has an interrupt output and two priority pins, priority in and priority out. The priority in pin on the highest-priority peripheral is tied to the active state. The priority out of that peripheral is tied to priority in of the next-lowest-priority peripheral, and so on. A common open-collector or open-drain interrupt line is driven by each peripheral.

If any peripheral is not requesting an interrupt, its priority out follows priority in. If a peripheral is requesting an interrupt, priority in is blocked. A peripheral can assert the interrupt request only if its priority in is active.

The result of this is that the highest-priority device in the chain that wants to assert an interrupt will do so. Lower-priority devices that want to interrupt the processor must wait until all higher-priority devices have been serviced. When the CPU responds to the interrupt request with an acknowledge cycle, the requesting peripheral provides the interrupt vector.

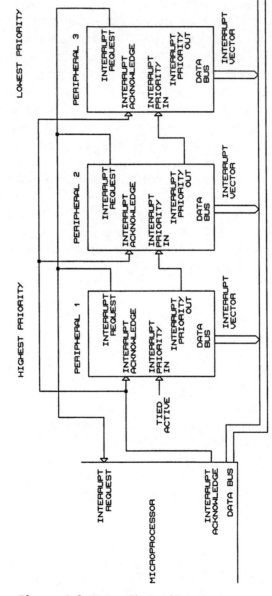

Figure 6.8 Daisy-Chained Interrupts

The interrupt priority of each peripheral in a daisy-chained system is fixed and is determined by the position on the chain. The primary advantage of daisy-chaining is that there is no limit to the number of peripherals that can interrupt the processor. Disadvantages include the fact that interrupt priority is fixed, and the potential that a low-priority peripheral may not be serviced in a timely manner, or even at all.

The Motorola 68000-family processors add another feature to the daisy-chained structure. Motorola has three priority-encoded interrupt requests. A peripheral requests an interrupt by driving these lines to the requested priority level. If the priority lines contain a binary code greater than the current state of the processor, the interrupt will be serviced. The priority lines allow two levels of prioritization—the priority inherent in each daisy chain, and an overall priority level in multiple daisy chains. Or, the priority levels can be used by individual, non-daisy-chained peripherals.

If a daisy-chained peripheral is not being serviced, the first thing to do is look at the priority in. If this signal is always inactive, some higher-priority device is holding the priority off. This may be because the peripheral never received a valid EOI command after a previous interrupt.

If a daisy-chained peripheral gets serviced, but too late, you can look at higher-level priority out signals on a DSO or logic analyzer to see which higher-priority peripheral is holding things off too long. Of course, if you have a general throughput problem that prevents the peripheral from being serviced, you will not find a single priority out line that is holding things up.

In a 68000-type system, if things hang up, look at the encoded priority lines. If they are stuck in some state (or if a single line is stuck), the processor state is probably preventing the interrupts from being serviced. Look for an ISR or synchronization error that left the processor priority in the wrong state.

Shared Edge-Sensitive Interrupts

Sometimes it is necessary to have two devices share a single edge-sensitive interrupt. Although this should generally be avoided, you might have a situation where an interrupt controller (such as the 8259) requires that all interrupt inputs be edge- or level-sensitive (not individually programmable) and one or more peripherals requires an edge-sensitive interrupt.

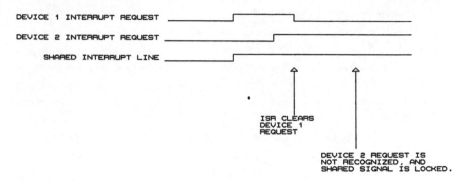

Figure 6.9 Shared Edge-Sensitive Interrupt Lockup

Figure 6.9 shows how a shared edge-sensitive interrupt might become locked. Two devices, device 1 and device 2, share a single edge-sensitive interrupt line. The two devices might drive the line through an OR gate or through a wire-OR (open-collector) scheme.

In the figure, device 1 asserts the interrupt. While the software is servicing device 1, and before the interrupt condition is cleared, device 2 asserts an interrupt. Since there was no transition on the interrupt line, the interrupt controller does not recognize the second interrupt.

The fix for this is to have the ISR check for other interrupt request conditions *after* clearing the first interrupt. An alternative solution is to provide a gate in the common interrupt line so that the software can momentarily force the common interrupt line to the inactive state after clearing the first interrupt and sending the EOI to the controller. This will produce an inactive/active transition if a second device has requested an interrupt.

Missing Interrupts

Missing interrupts can take the form of missing data from a UART, from a motor that runs away because of missing encoder interrupts, or even from a system that crashes.

If you are using an external controller, you can sometimes catch a missing interrupt by connecting a logic analyzer to trigger when there are two interrupts without an interrupt acknowledge cycle between them. This works only if you have a single-interrupt system, or if you can set the analyzer up to determine which interrupt is being acknowledged.

If you are using edge-sensitive interrupts, toggle a port bit at the beginning of the ISR. If you see two interrupts without seeing the port bit between them, you've missed one.

A similar trigger can be accomplished by outputting an action code at the start of the ISR. Again, set the analyzer to trigger on two interrupts with no start-of-ISR action code between them. This will probably require a logic analyzer with the ability to mix state and timing triggers.

An Interrupt Status Circuit

In debugging interrupts, it is often useful to have a real-time status indication of which ISRs are in progress. This allows you, using a logic analyzer, to correlate the state of the interrupts with other system conditions (such as an error indication). As discussed earlier, if you are using a microcontroller, you can set a port pin at the start of the ISR and clear it at the end of the ISR.

If you are not using a microcontroller, or if you are using a microcontroller but have no spare port pins, the circuit in Figure 6.10 may be a useful tool.

The circuit shown in the figure has four 74xx74 ICs wired as set/reset flipflops. Each flipflop represents one interrupt and is set or cleared by an output of a 74xx138 (3-to-8 decoder). The lower address lines from the CPU are used to select which flipflop will be set or cleared. The –STROBE input to the 74xx138, which enables the outputs, can be generated from a decoded I/O or memory select. The effect is to provide four bit-addressable port lines for indicating which ISRs are active.

Of course, the interrupt status could be stored in an 8-bit register. The advantage to this discrete flipflop design is that there is no need to maintain a mask of the register bits and no risk of race conditions while the mask and register are being updated. In operation, an ISR will, on entry, access the address that sets its flipflop. On exit, the flipflop is cleared. Neither operation affects the other flipflops in the circuit.

Also shown in the figure are two other methods of generating the –STROBE input to the 74xx138. The write-to-ROM circuit generates a strobe when writing to the ROM address space, and is suitable for use on processors with Intel-style read/write strobes. As with the write-to-ROM method described for action codes in an earlier chapter, make sure the ROM outputs are not enabled. The associated table shows which addresses generate which actions using this method.

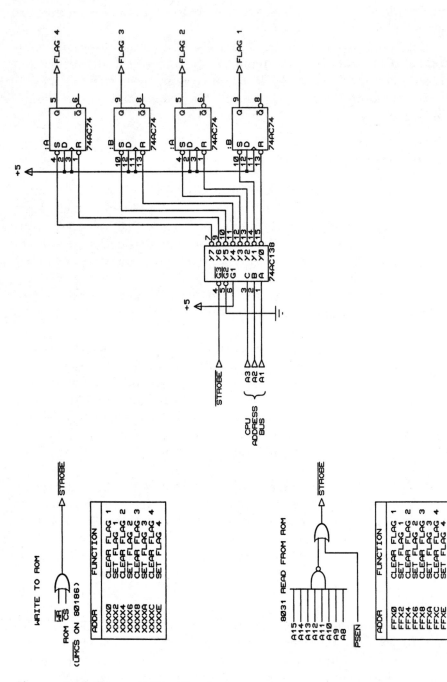

Figure 6.10 Interrupt Flag Circuit

The second enable circuit shown in the figure is suitable for an 8031-type processor or other Harvard-architecture processor that will not perform a write-to-ROM operation. In this case, a read from the upper 256 bytes of the ROM space generates the –STROBE signal. Again, the table shows which functions go with which addresses.

Although the circuit is shown with discrete logic, I would implement it in a PLD, which would also provide more than four outputs. When needed, the circuit could be clipped to the EPROM or connected to a header that brought out the required signals. The software can generate the set/clear strobes whether the circuit is connected or not, so no special version is needed for debugging.

Finally, when designing an interrupt system, remember the two immutable laws of interrupts:

1. An interrupt can occur at any time, between any two machine-level instructions.
2. All interrupts take time to process.

Most of the problems illustrated in this chapter can be prevented by remembering the following rules:

Rule 1: Avoid any variable (hardware or software) that is written by both ISR and non-ISR code. Variables (and hardware registers) should be written by the ISR and read by the non-ISR code or vice versa.

Rule 2: Where shared variables (or hardware registers) cannot be avoided, identify and protect all non-ISR (and interruptible ISR) code that must be indivisible.

Rule 3: Always assume that ISR-modified variables will change between two successive reads.

Rule 4: In any system with multiple interrupts, at some point in time, all the interrupts will stack up. Count on it.

Rule 5: The real world keeps happening while interrupts are being serviced.

Rule 7: Keep ISRs as short and simple as possible.

All the rules for non-ISR code apply to the ISR code as well, if nested interrupts are used.

7

*Debugging in an RTOS Environment**

Cody dropped into Matt's chair. "We finally released the programmer," he said. "Purchasing is buying parts and production has all the drawings." He pointed to the mounted circuit board on Matt's desk. "What's the story behind this? I looked up the part number that's silk-screened on the board, but I couldn't find any drawings or schematics. Is it some kind of top secret project or something?"

"No," Matt replied. "It was never released."

"So why keep it around?"

"As a reminder."

"So are you going to tell me the story, or just let me make something up on my own?"

Matt reached around and clicked his mouse, closing the open window on his screen, then swiveled to face Cody. "I came in late on that project. It was behind schedule and the projected costs were way too high and the software was full of bugs."

"Sounds like the ClearScan project you told me about."

"This was an option package for ClearScan. It originally consisted of three boards, each with a microprocessor and A to D converters, and just a lot of stuff. There was this meeting with Josh, and his boss, and the VP of marketing. I was told that if I would postpone my vacation, get this thing back on track, in budget, under cost, and generally fix the problems, I'd be a hero. They said they could 'see me getting a big bonus' for putting in the extra effort. The marketing guy said he'd pay my way on the annual marketing boondoggle to the Caymans if I could pull it off."

**Note:* In several places, this chapter uses as an example the RTXC kernel from Embedded System Products, whose people were very helpful in providing information about their product.

"Did you?"

"Sort of. I told my kids we'd have to postpone Disneyland for a while, and I worked evenings and weekends for three months. I compressed the design from three boards to one, completely rewrote the software, debugged everything, and got it working a week ahead of the deadline."

"So what happened then?"

"The ClearScan problems had been brewing all along, and things finally blew up. Everyone was backing away from the project, trying to shift responsibility for it, and Josh called me in one day and told me to archive the documentation because we wouldn't be releasing any ClearScan options for the foreseeable future."

"How about the bonus and the trip to Cayman?"

"Never heard another word about it."

"Were you mad?"

"Not yet. I went back to the project I had been on, which was now behind schedule. They had assigned someone to take over what I was doing, but that person had his own workload and never really got to my stuff. I told my kids we'd try to fit the trip to Disneyland in before school started, and went in the next day to tell Josh that I was going to take my postponed vacation. He said that we needed to get the current project caught up first. By the time that was done, the kids were back in school. When school was almost out, I told the kids we'd try Disneyland that summer. The oldest one said he wasn't interested, and I realized he'd outgrown it in a year. The younger one asked me if I really meant it this time. I told her I did, and scheduled the vacation."

"Did Josh let you go?"

"He came in a week before I was supposed to leave and said he needed for me to postpone my vacation again that year. I laughed out loud, pointed to the board, and he nodded and walked out. He hasn't asked since. I keep the board here to be sure I don't forget what's important."

Cody blinked. "Doesn't that sort of thing keep you from getting ahead?" he asked.

"Depends on what you mean by getting ahead. I might have missed a promotion or two. But I suspect there's never been a person since the Industrial Revolution who ever lay on his deathbed wishing he'd spent more time at the office. I think I'm getting ahead where it counts." Matt looked at his watch. "It's about time to go. You leaving?"

"Not yet. I've got to get a design specification done. Josh wants to see it first thing in the morning."

"How long do you think it will take?"

"Couple of hours."

"Tomorrow's Friday. What will Josh do first thing in the morning?"

Cody thought. "The weekly staff meeting?"

"Right. The one that usually lasts until about eleven. And then I heard him say he's going to the dentist. He won't be back until after lunch. If you wait until tomorrow, you've still got half a day to do the spec."

Cody sighed. "Does he do this sort of thing on purpose?"

Matt shook his head. "He's just going from day to day, like most people. He says he wants something first thing, but he forgets that he's got a meeting or a dentist appointment, and when he remembers he has other things that are more important than letting you know."

Cody nodded. "Give me a minute to get my briefcase and I'll walk out with you," he said.

Real-Time Operating System Functions

A *real-time operating system* or *real-time kernel* provides certain system management functions for an embedded system. The kernel (sometimes called a *real-time executive*) acts as a sort of middleman between tasks. If the programmer used a kernel, for example, the serial receive task might tell the kernel "Here is some data from the host PC. Wake up the receive data processing task and give it the data."

Different kernels provide different services and features. A full operating system will typically provide file and I/O management, and contains a kernel at its core. The basic kernel functions are as follows.

Task Scheduling

When a kernel is used, the software is divided and grouped into tasks. The RTOS provides services that permit a task to run, to block another task, to wait for some event before running, and other task-control functions. The purpose of task scheduling, depending on how the software is structured, is usually to do one or more of the following:

Make sure the high-priority tasks get done in a timely manner.

Make sure all tasks get to run sometimes.

Divide the available processing time between all tasks (but not necessarily equally).

Schedule tasks to run based on some event (such as the passage of time).

Communication

The kernel provides standard communication mechanisms and manages messages passed between tasks. Different kernels provide different capabilities, but some form of the following are typical:

> Mailboxes permit many tasks to send messages to a single task.
>
> Buffers or queues permit data to be passed between tasks.
>
> Semaphores permit binary indications to be passed between tasks. Some kernels provide more sophisticated semaphore functions, but they all provide at least the basic on/off functionality.

Memory Management

In a non-kernel-based system, memory that is needed will be allocated by the programmer (in assembly) or sometimes by the compiler (in an HLL) when the code is compiled. There are two types of memory allocation in kernel-based systems. *Fixed memory allocation* requires that all memory be allocated at compile time. *Dynamic memory allocation* permits memory to be allocated by the kernel at runtime.

When dynamic allocation is used, a task that needs memory, such as a buffer to put received serial data into, must request the memory from the kernel. The kernel will allocate the memory, if it is available, and keep track of which tasks have which memory. If hardware-based memory management is available, the kernel can also force a berserk task to stay out of memory belonging to other tasks (if support for the hardware is provided in the kernel).

Resource Management

As with memory management, the kernel may provide services that control access to other system resources, such as a printer interface.

A kernel provides a level of organization to a complex embedded application that a polling loop or pure interrupt-driven system does not. The kernel functions provide a common interface for communication and centralize the task of managing system resources. Although using a kernel makes some mistakes nearly impossible to make, it also adds an extra level of complexity, and therefore additional opportunities for other errors.

Preemption

Most kernels support several methods of ordering the priority of tasks. Many programs use roundrobin scheduling, where each task is tested in one big loop to see whether they can run. The PROM programmer uses roundrobin scheduling.

Most systems that are designed with a kernel use *preemptive scheduling*. In this scheduling method, the highest-priority task that can run gets control of the CPU and executes until it is finished or cannot proceed further (usually because it is waiting for something). When the highest-priority task cannot run, the second-highest priority task that can run gets control, and so on. In this chapter, we will assume the use of preemptive scheduling, except where otherwise noted.

Tasks

To manage task execution, the real-time kernel must know the following things about each task. This information is typically maintained in the Task Control Block (TCB):

Identifier: A number that identifies the task to the system.

Priority: The priority of the task relative to other tasks. Some kernels require all tasks to have different priorities; others permit two or more tasks to share the same priority.

Address: The entry point for the task code.

Stack: At initialization, the kernel must know the required stack size. Thereafter, the address of the top of the stack for each task is kept in the TCB.

State: The current state of the task must be known. The task can be ready to run, or it can be blocked.

Task State

The task state deserves a closer look. As already mentioned, a task can be ready to execute or it can be blocked. A task that is ready to execute will be allowed to run as soon as it becomes the highest-priority task. Put another way, a task that is ready to execute but never becomes the highest-priority task will never run, just as if it is blocked.

A blocked task is not ready to run because it is waiting for some external event. These events can include:

The task is inactive. For instance, if the programmer used a kernel, as described before, the receive data processing task might remain inactive until data was available. When the UART handling task received data, it would activate the receive data processing task.

The task could be waiting for a queue to contain data. Going back to the programmer example, the receive data processing task might not go inactive when it had processed all the data. It might instead just be blocked until the input queue (the Rx FIFO in the actual code) contained data, which would have been placed there by the UART handling task.

The task could be waiting for a semaphore or other data from another task.

Time delay: The task could be delayed for a specific time interval.

Resource availability: The task could be blocked because a specific resource is not available. This could be a hardware resource, such as if a task wanted to send data via the programmer serial port, but the port was already being used to send data for another task. The resource could also be memory, if the task has requested a memory block, but the amount of memory requested is not free.

The kernel manages tasks by maintaining a list of tasks that are ready and executing each one as it becomes the highest-priority task.

Use of a kernel will usually simplify the design of a system, since it adds a standard interface and many of the task-scheduling functions are built in and already tested. The kernel does, however, increase the system complexity because it adds a layer of oversight that wasn't there before.

An analogy of this is the document-control procedures for a small company in the days before CAD. Initially, the schematics and drawings are kept in a file cabinet, and engineering or production or sales can get to the documents. As the company grows, problems are encountered as engineering marks up an original drawing, or someone makes a mistake because someone else had revision G of a drawing checked out, so the first person was working with revision F because that was the latest version in the drawer.

Eventually, the company institutes a document-control department. People who want drawings must issue a request, and the people in document control make sure that only the latest copies go out. The problems where two people were trying to make changes to the same drawing are eliminated, because document control makes sure that change requests are processed in an orderly fashion.

The drawbacks are that everyone now has to go through the document-control process to get any drawings. Engineering may want drawings *right now*, but requests from production are higher priority, so they go to the head of the line. Engineering has to wait, unless they can convince whoever is in charge that their request is more important.

Systems based on a real-time kernel are susceptible to all the problems that any other real-time system has. Use of a real-time kernel, although it organizes and structures program execution, also adds characteristics that make debugging more difficult.

For an example of this, we'll look at the PROM programmer software. The programmer serial transmit code places data into a FIFO buffer. The transmit ISR sends data until the FIFO goes not empty.

In a kernel-based system, the receive processing code could be structured this way:

Receive ISR, which sets a semaphore (via the kernel) to indicate that a character is available.

Receive input task, which waits for the data available semaphore (from the receive ISR) and then reads a byte of data. The receive input task waits until an entire line has been received (terminated by the carriage-return character), then passes the data to the receive processing task via a kernel-managed queue.

Receive processing task, which processes the input commands and data.

Now suppose that we have discovered a problem in our system where a byte is being lost during hex downloads, detected because the end-of-line checksum is incorrect. We don't know if the data is lost because the receiver overran, if the receive input task threw the byte away for some reason, or if the receive processing task failed to store the data in the device buffer for some reason.

In a non-kernel based system, we could look at the input FIFO (which is at a fixed location in memory) when the error is detected, and we would know whether the byte was there or not. This would draw a binary line between the receive ISR and receive processing functions (receive ISR and receive input are common).

In the kernel-based design, using dynamic memory allocation, looking at the data that is output from the receive input task is more problematic. The queue used to pass data from receive input to receive processing is dynamically allocated by the kernel. It is not at a constant address or even constrained to a small segment of the RAM the way the FIFO buffer is. Consequently, it is difficult to look at the queue with normal tools. It is not impossible to do so; the internal parameters can be deciphered to determine where the queue is located. This is a tedious job, however.

A better solution is to use a debugging tool that is aware of the RTOS and can look directly at the queues, semaphores, and mailboxes.

When designing with an RTOS, the use of RTOS-aware tools greatly simplifies the debugging process.

Some kernels do not dynamically allocate queues, but instead require them to be defined at initialization. These kernels would not have the debugging problem described, since locations of all the queues would be known. The same problem might exist, however, for dynamically allocated semaphores.

Priority

In any preemptive system, the highest-priority ready task will always run, and it will run until it finishes or becomes blocked. This means that, as long as the highest-priority task is ready to run, no other tasks will run. To avoid having tasks blocked that need to run, make sure the high-priority task contains only high-priority actions. Lower-priority actions should be in a lower-priority task.

Let's look at the programmer for an example of how this can be a problem. Say that we have written a programming procedure for a new type of IC. Let's say that we have to drive a programming signal to the active state, and to avoid burning up the target PROM, we must turn the signal off within 100 microseconds. If a receive interrupt occurs after the signal is turned on, the receive command processor may get control and cause us to exceed the maximum time limit. Normally, it would make sense for the receive command processing to be a higher priority than the programming function (so we can issue a command to stop programming, for instance), but in this case that may cause a timing problem. This is not an insurmountable problem, and could be solved by making different command functions different priority (some lower than the programming function) or disabling interrupts during critical sections of the programming code (but not long enough to cause data to be lost). The point is that avoiding this sort of problem requires you to adjust your thinking when using a kernel.

Interrupts in a Kernel Environment

Interrupts are as necessary when using a kernel as they are for a non-kernel design. All kernels require a regular tick interrupt, which is used for timekeeping and for allowing the kernel to have periodic control of the CPU. This is in addition to whatever other device interrupts the system requires.

Using a kernel forces interrupts to be handled differently than in a system without a kernel. Using RTXC as an example, there are three key rules for interrupts:

1. The interrupt vector does not point to the ISR itself, but points instead to a kernel routine that performs a task switch to the ISR.
2. The ISR cannot call kernel services, except for specific ISR-related functions. This is because the kernel services are not reentrant, and the interrupt can occur during kernel execution.
3. The ISR does not execute a normal return instruction, but instead exits through the kernel using a specific kernel service. The kernel can then switch to a higher-priority task or perform other task-management functions as required.

To see what impact this structure has on the design and execution of code, we'll look at the programmer UART output function.

In the code as it is presently structured, the UART transmitter generates an interrupt whenever the transmit data buffer is empty and when the interrupt enable (IE) bit is set in the UART.

A task that wants to transmit data via the UART places the data into the transmit FIFO and sets the IE bit. When the UART interrupt occurs, the UART ISR finds the transmit interrupt bit set, transfers a byte from the transmit FIFO to the UART, and resets the IE bit if it were the last byte (FIFO empty).

In the RTOS environment, the sequence is different. The RTXC kernel assumes that the ISR will turn off the interrupt request, but the ISR cannot transmit data (because it cannot use kernel services to get the data), so the process goes something like this:

A task that wants to transmit puts data in a mailbox or queue via the kernel.

The UART IE bit gets set, probably as a result of a task activated by data in the queue.

The UART ISR detects the transmit interrupt and disables the UART IE bit. It returns via the kernel, using an ISR service to set a semaphore, which we'll call TXREADY.

A task, which we'll call UARTOUT, has been blocked waiting for TXREADY to be set. The kernel sets TXREADY in response to the

ISR request and then unblocks UARTOUT. UARTOUT resets TXREADY (via the kernel), gets a byte from the transmit queue or mailbox, and transmits it. If the queue or mailbox is not empty, the UART IE bit is set, and then the UARTOUT task blocks again, waiting for TXREADY.

Obviously there is more overhead in the kernel-based design. In the non-kernel design, the UART IE bit is set to start transmission, and is cleared when the last byte is sent. The kernel assumes that the ISR clears the source of the interrupt, so it reenables the interrupt. If the ISR did not clear the UART IE bit, the interrupt would be continuous. In the non-kernel design, the ISR would transmit the data, clearing the interrupt without needing to clear UART IE. Since the kernel-based design cannot do this, it must clear UART IE and set the semaphore that kicks off UARTOUT. UARTOUT must transmit the byte and set the UART IE bit to enable the next interrupt. This method works, but there is obviously a lot of setting and clearing of the UART IE bit.

Some kernels are more flexible in letting ISRs perform task-like kernel function calls, but at the expense of more overhead.

Priority Inversion

This is a known problem with real-time kernels, or, more specifically, with any preemptive structure.

Let's say that the programmer is implemented with a preemptive kernel. The following tasks are involved in this example, and are listed in order of priority:

Command Processor (processes commands from the host PC).

Stop Key (monitors stop pushbutton every 50 ms using the kernel timer service).

Send Msg (transfers message to RS-232 output process).

Let's say that the Command Processor is blocked, waiting for input data. Since there is only one transmit line back to the PC, we have structured Send Msg to use a semaphore called GOTSENDER. When any task wants to send something via the RS-232 port, it must wait until GOTSENDER is cleared. This prevents a high-priority task from inserting a message halfway through transmission of a message from a low-priority task.

The operator has commanded the programmer to receive an Intel-format hex file, but the file he sent is a raw hex file. He discovers this and presses the Stop Key. While our Command Processor is blocked, the 50 ms timer expires and Stop Key gets control (it is the highest-priority task that can run). The timer detects that the operator pressed the stop pushbutton, and, via the kernel, sets GOTSENDER to indicate possession of the Send Msg resource. The pushbutton code then proceeds to transfer a message via Send Msg.

Halfway through sending the message, another character is received, and the Command Processor gets control again. Finding that the character is invalid, the Command Processor wants to send an error message to the host PC. However, when Command Processor requests control via GOTSENDER, it blocks because the serial output function is not available. The kernel now transfers control back to the Stop Key code (once again the highest priority runnable task). Stop Key completes transferring the data, releases GOTSENDER, and the Command Processor now gets to transmit.

Priority inversion occurred here because the high-priority task (Command Processor) was effectively forced to run at the priority of a lower-priority task (Stop Key) because the lower-priority task had a resource that the high-priority task needed (GOTSENDER).

Most examples of priority inversion illustrate that it can be a problem. In this case, priority inversion is actually beneficial: if Command Processor had been able to get control of the Send Msg resource, the user would have seen half the Stop Key message, then the Command Processor message, then the rest of the Stop Key message.

The key components of priority inversion are:

- A high-priority task is forced to wait for a lower-priority task . . .
- because a lower-priority task has a resource the higher-priority task needs.

Priority inversion can be very difficult to find when it involves three tasks, as follows:

- A high-priority task, HIGH, is waiting for a timer to expire.
- A low-priority task, LOW, gets control and takes a semaphore.
- HIGH's timer expires and HIGH preempts LOW. HIGH attempts to get the semaphore that LOW has, and blocks because the semaphore is unavailable. Meanwhile, some event causes a medium-priority task, MEDIUM, to be ready to run.

Since HIGH is blocked, MEDIUM gets control. MEDIUM has a higher priority than LOW, so it runs until completion. HIGH effectively has the same priority as LOW at this point.

There are a number of ways to prevent priority inversion:

- Allow the low-priority task to change to a high-priority as long as it has control of any resource that is shared with a higher-priority task. This is only feasible, of course, if the kernel permits tasks to change their priority. Using this method requires that the programmer keep track of all resources that are shared with a higher-priority task. If the code for a high-priority task is changed so that it uses semaphores it didn't use before, all lower-priority tasks must be examined to see whether they use that semaphore.
- Some operating systems have mechanisms for preventing priority inversion. RTXC, for example, allows a flag to be set that will permit the kernel to check for priority inversion.
- Some kernels allow a task to see whether another task has control of the resource. This permits the task to proceed, coming back to the unavailable resource later.
- Some kernels (RTXC is one) permit a task to wait on a semaphore with a time-out. If the semaphore does not become available in a specified time, control returns to the task anyway (provided, of course, that a higher priority task has not preempted it).

General Preemption Problems

The use of a preemptive kernel introduces certain potential errors that do not exist in a roundrobin or other scheduling mechanism.

Let's look at the programmer code again. In the roundrobin scheme the programmer actually uses, the Command Processor executes commands as they are decoded. One of these is the command to fill the programming buffer with a constant value.

If the user sends a script file that strings several commands together back-to-back, the Fill command can be followed immediately by a command to download Intel-format hex data. The programmer will recognize the Fill command and begin to execute the fill operation. Fill fills memory starting from the lowest address.

As the Download command is received, the UART will interrupt the Fill operation. The UART receive ISR will buffer the incoming

command in the receive buffer, and control will return to the fill code
after each byte is received. After the Fill operation is complete, control
is returned to the background loop, the Command Processor will get
control again, and the Download command will be executed.

In a kernel-based system, the programmer might structure the
following tasks, in the following order of priority:

Command Processor

Download

Fill

Now let's see what happens: Command Processor gets the Fill
command and tells the kernel to initiate the fill task.

As Fill is executing, the Download command is received as
before. This time, however, the Command Processor gets control every
time a byte is received via the UART ISR. When the complete Download command has been received, the Command Processor tells the
kernel to start the download task.

The download task preempts the fill task every time a byte is
received. The result may be that the download task will put data in
memory that is then overwritten by the fill task. This happens because
Intel-format hex files are not required to load data in any particular
order. Since each line of data has a unique starting address, the data
can be loaded in any order, including the top segment of memory first.

Since *most* Intel-format files have the data in ascending order, the
fill task will usually stay ahead of the download task and no harm is
done. But this may cause the problem to appear months later, in the
field, and only when a particular user codes things a certain way.

The fixes for this problem are simple:

Make the buffer a resource that is locked with a semaphore so the
Download cannot write to it until Fill is done.

Make Fill higher priority than Download.

Add code to the Command Processor so that Download is not
started until Fill is finished.

This problem is really an interrupt-caused race condition, as
described in the chapter on debugging interrupts. The event that
causes the race is the UART receive interrupt. As illustrated, the fixes
for this particular problem are straightforward. The problem itself,

however, is typical of problems that can be introduced by using a kernel. The kernel adds a level of abstraction to the code that can make situations such as this more difficult to spot.

Use of a preemptive kernel can also affect communication between tasks. Any call to the kernel can result in a task switch if a semaphore is set, or if a message is sent (or received), or even if an ISR occurs.

As described in an earlier chapter, the programmer has a function called ASCHEX that converts ASCII data in ASC to hex data in HEX. In the polled environment, if a task, such as the receive Command Processor, needs to convert ASCII data to hex, it places the data in ASC, calls ASCHEX, and processes the returned data in HEX. When the Command Processor transfers control to ASCHEX, via a subroutine call, no other code except that in ASCHEX can run (except for ISRs) until ASCHEX is finished and has returned.

In a kernel environment, the functionality would be somewhat different. ASCHEX might be a task, activated when a message (containing the ASCII data) is received. The Command Processor sends a message containing the ASCII data to ASCHEX, then blocks, waiting for the results.

ASCHEX has been blocked, waiting for a message. The kernel activates ASCHEX, which reads the data, does the conversion, and sends a message back to the Command Processor with the converted data in it. The kernel unblocks the Command Processor, which does whatever it wants to do with the data.

This is a simple scenario, but it contains some potential pitfalls that do not exist in the polled environment.

The first potential problem is in the ASCHEX function itself. In the polled environment, ASCHEX can do its thing and return when finished. In the kernel environment, ASCHEX must make its return message the last thing it does, because the act of passing the return data message, via the kernel, will reactivate the command processor. Any processing that ASCHEX has left to do (cleanup of buffers or whatever) won't get done, at least until ASCHEX is activated again.

The second problem involves the fact of preemption. In the polled environment, the Command Processor calls ASCHEX and resumes execution when ASCHEX returns. In a kernel environment, the activation of ASCHEX (by passing a message) and reactivation of the Command Processor (by the return message) are both operations performed by the kernel. Every kernel operation has the potential to cause preemption.

Say that the Command Processor is a very high priority task and ASCHEX is very low. Suppose that when the Command Processor sends ASCII data to ASCHEX, there is a task of intermediate priority ready to run. As soon as the Command Processor makes the kernel call to pass the message, the kernel will block the Command Processor (which is now waiting for return data), unblock ASCHEX, and then transfer control to the intermediate task. Since the intermediate task has a higher priority than ASCHEX, it gets control first. It may be some time before control finally passes to ASCHEX, and then back to the Command Processor.

This situation is a problem only if it is unexpected and causes the system to malfunction. More importantly, it is a problem that does not occur in the polled environment because the polled code does not use preemption.

Deadlock

Deadlock is defined as a condition where a task locks up permanently because some other task has a resource it needs. Typically, the second task will not release the resource until the first task proceeds, and the first task will not proceed until the resource is available.

As an example, take a look at the fill/download problem again. Let's say that we changed the code so that the memory buffer is treated as a resource, with a semaphore to determine the availability of the buffer. The download and fill functions are still separate tasks, but to prevent overwriting download data with the fill constant, Fill now sets a semaphore when it is finished. Download will be activated by the command processor, and will request the memory buffer, but it blocks unless Fill is done.

Now let's say that we have a bug in our code so that Fill, instead of requesting the buffer and keeping it until done, requests the buffer from the kernel for every byte it wants to store. The sequence of events is now:

Fill is busy filling the buffer, requesting and giving up the buffer for every byte.

Command Processor decodes the Download command activates the download task.

Download requests the buffer and acquires it. Since Fill has not yet set the semaphore to say it is done, Download now blocks, waiting for Fill to finish.

Fill will never finish because Download has possession of the buffer.

Although this is a contrived example, it illustrates how deadlock can work. In this case, the simple fix is for Fill to request the buffer once, giving it up when done. Download could then wait for the buffer to become available, eliminating the separate "fill done" semaphore.

Data Transfers

Passing data between tasks in a polled environment is straightforward. You put data in a buffer of some kind and either call the recipient directly if it is a subroutine (such as ASCIIEX in the programmer) or set a semaphore. If the semaphore is tested in the polling loop, you know the recipient will get the data on the next pass through the loop.

Communication in a kernel environment is somewhat different. Data is typically placed in a FIFO queue or in a mailbox and passed to the recipient through the kernel. This scheme provides several advantages, including the use of a standard, common interface.

There some drawbacks to data transfers when using a kernel:

- If the recipient is blocked, it may never see the message.
- Some kernels, such as RTXC, permit messages to be assigned a priority. If the recipient has a continuous stream of higher-priority messages, it may never get to a low-priority message.
- Some kernels (again, RTXC) permit a task to look at messages only from a certain sender, which would cause it to miss messages from a different sender, even if those messages had higher priority.

Some of these problems can be avoided by allowing the sender to be notified when the recipient has taken the message or by providing a time-out to the sender if the recipient doesn't take the message (both features of RTXC). Some of these problems can show up in one form or another in a polled system as well, but the kernel adds a layer of complexity that may make the problems harder to see.

The potential problems illustrated so far are necessarily simple to make the concepts clear. However, in a real system, problems are more difficult to debug because tasks are constantly being blocked and unblocked, and messages are being passed back and forth continuously. As with a shy child in a room full of children at a noisy party, it is not always easy to see the forlorn, low-priority task that never gets unblocked or cannot get a message through. It is important to know that *any* kernel call and *any* interrupt (including the timer tick) may cause the current task to stop running in favor of a higher-priority task. Use of a kernel does not automatically introduce problems. It does require rethinking how the code is structured and requires some foresight to avoid the potential pitfalls.

Action Codes

Although the use of a kernel usually implies more sophisticated tools than are sometimes available in simpler systems, action codes, as described in Chapter 2, can still be a benefit to a kernel-based design. Typically, action codes would be used to track instances:

When a task is activated

When a task is blocked (and maybe why)

When a task is unblocked

When a task is finished

When a task acquires/gives up a resource

When the kernel tick interrupt occurs

When other ISRs occur

Figure 7.1 shows a simple flowchart of the code of the polling loop for the programmer. Although flowcharts are not normally used for kernel-based software, Figure 7.2 shows a simple method that can be used to diagram a kernel-based system at a high level. Figure 7.3 shows some of the programmer tasks as they might be implemented using a kernel.

Debugging Tools

RTXCbug, provided with the RTXC kernel, is an example of a typical command-based debug tool. To use RTXCbug, your system must support a console device, such as a PC connected to the target via an RS-232 serial port. The console interface is linked when the firmware is compiled/built. You can activate RTXCbug within your application by calling a specific RTXCbug function, or by entering an exclamation

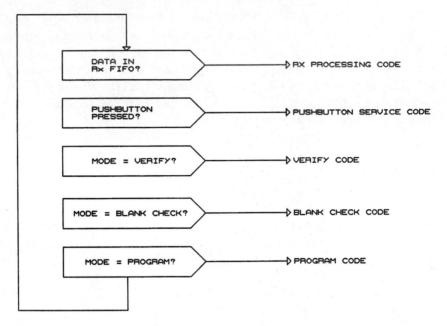

Figure 7.1 Programmer Polling Flowchart

point (!) on the console. For the second method, of course, the console interface must be able to generate an interrupt when a character is received.

RTXCbug executes as a kernel task, usually the highest-priority task. Consequently, other tasks cannot run while RTXCbug is active, although interrupts will be serviced. Using commands entered through the console, RTXCBUG provides a snapshot of the system at any given time. The commands available include:

Tasks: Displays the state of all the tasks, in the form

xx NAME yy STATE

where xx is the task number, NAME is the task name, yy is the priority, and STATE is the state of the task. The state can be INACTIVE, READY, –READY (blocked), DELAY (remaining delay shown), SUSPENDED, Semaphore (waiting on the specified semaphore), QueueEmpty (waiting because the indicated queue is empty), QueueFull (waiting because the specified queue is full), Mailbox (waiting because the specified mailbox is empty),

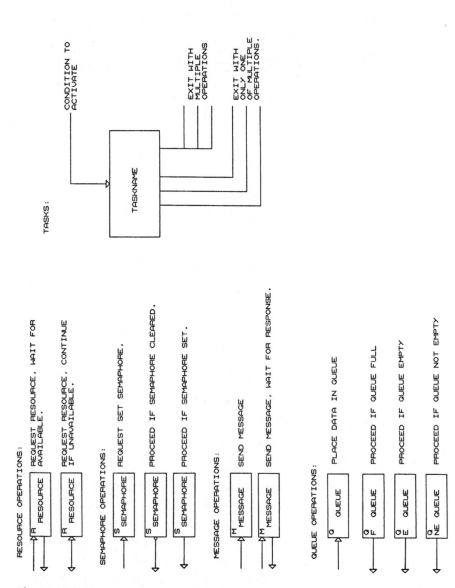

Figure 7.2 Programmer Diagramming Method

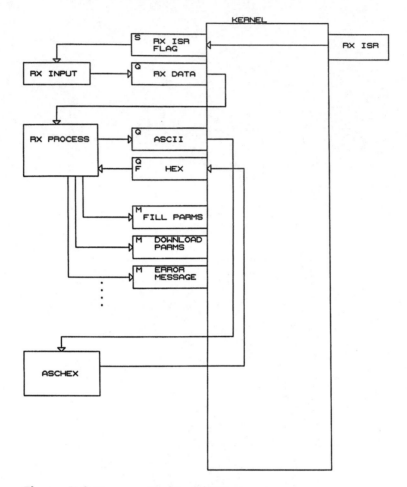

Figure 7.3 Programmer Kernel Diagram

Resource (waiting because the specified resource is locked), and
Partition (the specified partition is empty).

In all cases, if there is a time delay associated with a waiting task
(such as if the task is waiting for a mailbox to go not empty, but
with a time-out), the remaining time will be displayed adjacent to
the state. The name of queues, mailboxes, resources, partitions,
and semaphores are also displayed next to the state description.

Queues produces a list of the queues in the system, in the form:

xx NAME y/z worst count waiters

xx is the queue number, and NAME is the queue name. Y is the current size of the queue, z is the maximum depth. Worst is the worst case usage, count is the total number of items that have been placed in the queue, and waiters is the name of the tasks, if any, that are waiting for the queue.

Semaphores displays a list of semaphore numbers, names, states, and any tasks waiting on the semaphores.

Resources displays the resource numbers, names, a count of the times the resource has been locked and unlocked, a count of the conflicts (when the resource was requested but already in use), the task that currently owns each resource, and the names of any tasks that are waiting on the resource.

Other commands provide similar information about memory partitions, mailboxes, clock/timers, and stack space. A task manager function allows you to manually suspend, resume, and terminate tasks, change task priorities, and perform other task control functions.

Using the Debug Tools

You use a tool such as RTXCbug much like you would an emulator or a debugger. When a problem occurs, you invoke the debugging tool, either manually or by allowing your error handler code to call it. You examine the state of tasks, semaphores, memory, and other resources to determine just what went wrong.

RTXCbug allows you to debug and tune the operation of your code with respect to the kernel. However, it is not a debugger and does not provide you with a window to your application code, other than to see how the code interfaces to the kernel.

Kernel-Aware Debuggers

If you are going to use a kernel in an embedded application, a kernel-aware debugger is a significant asset. A normal debugger (or emulator) can trace execution of your code, but calls, task switches, and other kernel functions are just another block of code. You are not interested in tracing the internal operation of the kernel unless you think it has a bug. A kernel-aware debugger knows about the kernel and treats kernel functions as indivisible macro operations. Instead of displaying the machine-code instructions that are executed when the kernel switches tasks or sets a semaphore, the kernel-aware debugger just shows what operation was performed. This is a significant advantage to debugging in the RTOS environment. To make this effective, the compiler, RTOS, emulator (if used), and debugger must work together—not always easy to achieve.

In Chapter 8, we'll look at debugging embedded systems based on the personal computer architecture.

8

Debugging in an Embedded PC Environment

For various reasons, some real-time systems are designed around a standard computer platform, usually based on the IBM PC architecture. Although entire books have been written about designing and debugging software for the PC platform, this chapter concentrates primarily on those aspects that are unique to debugging real-time systems.

Advantages to Using a PC Platform

The first and most obvious advantage to using a PC platform as a basis for your design is the availability of tools. In a ROM-based design, you must use tools that support ROMable code. If you are using any processor other than x86, your choice of tools will be limited. On the other hand, if you design around the PC platform, then your real-time application will (probably) run under whatever operating system is running the rest of the computer, be that DOS or Windows. This means that every PC development language, from Basic to the latest object-oriented compiler, is an option.

In a ROM-based design, you will typically choose a processor that has sufficient capability to run your application. On a PC platform, you will probably use whatever the current processor technology is, and, since it must perform the normal PC duties as well as run your application, you often find that there is considerable throughput left over. This can be put to use running some of the advanced PC-based debugging tools that are available. Some of these include sophisticated performance monitors that can tell you where the software is spending its time and whether there is a potential throughput problem.

In a ROM-based design, you usually must debug your proprietary hardware before the software can be integrated. When a problem occurs, there is often a question as to whether the problem is in the software or is caused by an as-yet-undiscovered bug in the hardware timing. In the PC platform, your hardware is off-the-shelf and presumably working correctly. If you have not added any custom boards, any problems are almost certainly due to the software. This does not preclude, of course, software conflicts with added off-the-shelf hardware.

Because the target and the development system are often the same physical computer, the compile/run time is shorter since there is no time expended to program a PROM or download the code.

The hardware-based memory management in modern PCs provides no-cost protection to your application as well.

Disadvantages to Using a PC Platform

Although there are numerous advantages to using a PC or other standard computer architecture for an embedded system, there are a number of drawbacks as well.

Operating System

When you use a commercial real-time operating system, such as RTXC, you usually get a set of numbers that tell you how long it takes to switch tasks, how long interrupts will be turned off, and, in short, the things you need to know to make your software deterministic. Sometimes you get source code as well, allowing you to see exactly how the kernel performs it's functions. When using DOS or Windows, you are dealing with operating systems that are not designed to be real-time. You don't know how long interrupts may be disabled by the OS, and the source code is certainly not available. This makes debugging more difficult, especially if an intermittent problem is caused by the operating system sometimes taking a little longer than usual to perform some function.

Some of these problems can be alleviated by using an operating system that is designed for real-time applications. These products provide varying degrees of compatibility with the PC architecture, and varying degrees of timing predictability.

Uncontrolled Hardware

When using a PC platform, you typically buy off-the-shelf hardware for standard components such as the VGA video controller, hard drive controller, and so on. You rarely get schematics for these boards, and if you do, much of the functionality is embedded in ASICs so you don't know how it really works. The manufacturing life of these components is often very short, so there is no guarantee that what you buy this week will be available next month.

The software to support this hardware is often a problem as well. The BIOS for many VGA controllers, for example, leaves interrupts disabled for extended periods of time. Again, if you get your software working with one controller/BIOS combination, a software or hardware change from the vendor may introduce an entirely new set of problems.

This problem can be alleviated in many cases by purchasing the standard components from vendors that make PCs specifically for embedded applications and who build stable designs with significant product life. Such hardware usually costs more, however.

Interrupts

The PC-AT architecture allows for 16 hardware interrupts, using two cascaded 8259 interrupt controller ICs. In most modern motherboards, these controllers are embedded in a multifunction chip set, but the functionality is still there. The interrupt map for the PC-AT is as follows:

IRQ 0: 8253 Timer 0 (18.2 Hz)

IRQ 1: Keyboard

IRQ 2: Used to cascade second 8259

IRQ 3: COM 2 (if installed)

IRQ 4: COM 1

IRQ 5: Printer

IRQ 6: Floppy disk

IRQ 7: Printer

IRQ 8: CMOS real-time clock

IRQ 9: Redirected IRQ 2

IRQ 10: Unassigned

IRQ 11: Unassigned

IRQ 12: Unassigned (mouse on PS/2)

IRQ 13: Math coprocessor

IRQ 14: HDD

IRQ 15: Unassigned

IRQ 11 and 12 are often used for SCSI adapters. So, as can be seen, in a fully populated machine with two comm ports, there are only five unused interrupts (IRQ 5, 10 through 12, and 15). These can fill up quickly with off-the-shelf peripherals, leaving few choices for custom hardware that needs an interrupt to support the embedded application.

It is possible to share interrupts, but the PC architecture uses edge-sensitive interrupts, making this difficult. If interrupts must be shared, you must either:

- Make sure that only one device at a time uses the interrupt.
- Chain the ISRs, as described in Chapter 6, to be sure everything gets serviced.

Even if there are no hardware interrupt conflicts, the interrupt definition of the PC architecture conflicts with Intel's specifications for interrupt usage. This means that some interrupts are shared with machine exception conditions, such as stack overflow or memory management faults. Terminate-and-stay-resident (TSR) programs and hardware device drivers may intercept the interrupts, causing conflicts with your software. You must make sure your software is compatible with whatever other software gets loaded into the machine. If the customer adds your application to a general-purpose machine that is used for other things, it is difficult to predict all the interactions.

The last potential pitfall with interrupts involves latency. As already mentioned, normal PC operations sometimes leave interrupts off for unspecified amounts of time. This can wreak havoc on your program if it expects quick interrupt response. Even well-behaved BIOS functions can cause problems simply because of the number of interrupts necessary to run the system. Even with the latest, most souped-up processor, it is risky to use a PC for an embedded application that needs short latencies. If you have high-speed data that absolutely cannot be lost, consider adding a FIFO or other buffer in the design so that the interrupt latency doesn't break things. Measure the interrupt latency over a period of time, with the software performing various functions, to be sure it will not break when things get tight.

Throughput

A PC platform that controls an embedded application must handle all the normal PC functions such as keyboard, disks, video, and so on, as well as the embedded application itself. The system must have sufficient speed and memory to achieve all this without falling behind some-where. This means that the available throughput after performing all the PC-related functions must be sufficient not only for performing the embedded application, but for doing so in a real-time fashion. This often calls for a much more powerful processor than would be required for the embedded application alone.

Debug

A commercial real-time kernel usually has (or can be made to work with) kernel-aware tools. These provide a window into kernel-related functions such as queues, semaphores, and task status. DOS and Windows are not provided with such tools, making debug more difficult. If a crash occurs, the result is typically not a graceful exit through a breakpoint interrupt, but a DOS or Windows error message.

In an embedded PC environment, the application typically runs under DOS, Windows, or whatever operating system is used. This means that the disk, keyboard, display, and everything the OS needs to run is going in the background while you are trying to debug. Although you can single step your application code using OS-based debug tools, it is difficult it isolate a real-time application from the effects of the OS during debug. In addition, your application is typi-cally not the highest-priority activity for the CPU, which can lead to throughput problems.

Emulators

Emulators are rarely used for debugging embedded PC applications; they typically aren't even available for the highest-performance CPUs. Debug depends on OS-based tools.

Action Codes

Action codes can be used to assist in debugging an embedded PC application. If you have proprietary hardware in the system, and if you have room for a spare I/O decode on the board, you can create a debug port by adding a header to the top of the board. The header would include connections for ground, the decoded write strobe for the action codes, and the buffered PC data bus.

If you do not have a proprietary board, you can still use action codes by writing them to the printer port if it is otherwise unused.

Another possibility is to write the codes to a nonexistent I/O location, where they can be decoded by a separate board and plugged into the bus only when the debug information is required.

PC Debugging Tools

Although operating in a PC environment can make debug more difficult, there is an extensive library of debugging and performance monitoring tools available. A wide array of debuggers with powerful features simplifies the debug process. In addition, you can get programs that will monitor the performance of your program, telling you how long the code spends in a particular routine, what the interrupt latency is, and about anything else you want to know. Some of these tools can, in real time, produce a histogram of where the code is spending its time.

Photo 8.1 is a screen shot of PTach, a performance optimization utility from NewCon software, and is an example of the types of tools available.

In the next chapter, we'll look at some real-world debugging scenarios.

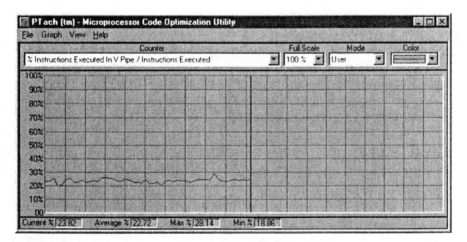

Photo 8.1 PTach Screen Shot (Courtesy NewCon Software, Pacific Grove, CA)

9

Debugging Scenarios from the Trenches

This chapter describes several real-world debugging scenarios. These are based more or less on real-world examples (some examples more, some less). Names and technical details have been changed to protect both the innocent and the guilty, and some details were left out, simplified, or altered simply due to the fallibility of human memory. Along the way, we'll poke a little fun at management and marketing, too.

A Debugging Scenario Using Action Codes

This first example is based very loosely on a real-world scenario, with the differences mostly to avoid cluttering the description up with confusing detail.

Figure 9.1 shows a 16550 UART used in a simple embedded system. The 16550 communicates with a host system via RS-232 and receives a continuous stream of messages. Each message has an opcode (1 byte), 2 bytes of data, and an 8-bit checksum.

The software is structured so that the UART receive ISR receives an entire message and notifies the background code via a semaphore when a message is received. Input data is buffered in an 8-byte FIFO, implemented in software. The RTS output of the 16550 is used to control data flow from the host. The receive ISR clears RTS to stop transmission when the FIFO holds 8 bytes. The background code reenables the host when the FIFO drops to 4 bytes.

We have calculated that the software, although it might not process a message immediately when notified by the ISR, is able on average to keep up with the data rate.

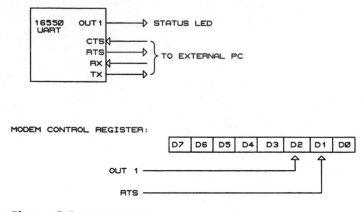

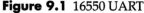

Figure 9.1 16550 UART

The 16550 has two user-defined outputs, OUT1 and OUT2, that are used by the background code to drive status LEDs. OUT1, OUT2, and RTS are all controlled by a single register, the Modem Control Register (MCR), in the 16550.

The non-ISR code to update the LEDs looks like this:
```
Read MCR
Set/clear LED bits
Write MCR
```

The ISR code to disable transmission looks like this:
```
Read MCR
Set D1 (RTS)
Write new value to MCR
```

As you might expect, the interrupt sometimes occurs between the time the unprotected non-ISR code reads the MCR and the time when it writes the updated value back to the MCR. The result is the ISR attempt to shut off the data flow with RTS is thwarted, since RTS is turned back on by the non-ISR code. The FIFO then overflows, and some data is lost. However, the error is buried in complex code.

Not yet knowing what the problem is, we see an intermittent symptom where some messages have checksum errors, caused by the missing data. We generate several possible theories to account for this problem:

The external system sometimes drops bytes.

The external system sometimes gets the checksum wrong.

The external system sometimes fails to recognize RTS.

The ISR doesn't recognize when FIFO is getting full, doesn't set RTS.

The ISR gets FIFO pointers messed up, doesn't know where it is.

The CPU sometimes calculates checksums incorrectly.

The code sometimes leaves interrupts off too long, UART over-runs.

Our assumption about average versus instantaneous throughput is wrong.

We use a protocol analyzer to look at the data and calculate several checksums by hand. We are able to absolve the external system of any blame, eliminating the first three theories.

We now add action codes to the software, writing them to an unused I/O port and capturing them on a logic analyzer. The codes (the ones we are interested in, anyway) are as follows:

80: UART ISR entry

81: UART ISR exit

83: 6 bytes in buffer—clear RTS

84: First byte of external message received

85: Last byte of external message received—semaphore set

20: Turn LED on

21: Turn LED off

46: Message processed by background code

E0: Checksum error detected

Our first list of action codes, captured on the logic analyzer, is shown in Figure 9.2. Notes have been added for clarification. This trace has been simplified; a real trace would be cluttered with other action codes as well.

Note that the problem actually occurs several bytes prior to the detection of the bad checksum. The symptom (bad checksum) is misleading because it is caused by a missing byte, caused by FIFO overflow, which in turn is caused by the actual timing problem.

BYTES IN
FIFO

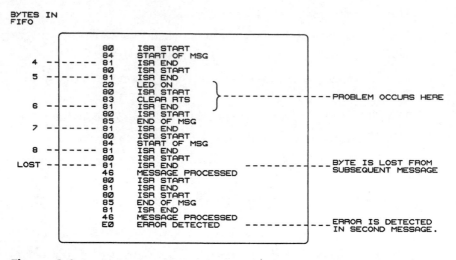

Figure 9.2 Logic Analyzer Trace for 16550 UART Problem

In running the system several times, we notice that we always have the following sequence in the bad message or in the preceding message:

20 (Turn LED on)

80 (Start of interrupt)

82 (Clear RTS)

81 (End of interrupt)

Suspicious that the LED might be somehow connected, we add an additional action code, 22, and place it immediately after the operation that actually turns the LED on. Now our sequence always looks like this:

20 (Turn LED on)

80 (Start of interrupt)

82 (Clear RTS)

81 (End of interrupt)

22 (LED turnon complete)

This narrows the problem down to a manageable section of code, which can be visually examined to find the cause. We could also trigger the analyzer on this sequence and use the trigger output of the analyzer to stop an emulator. A look at the trace buffer would show the two write operations that caused the problem.

This is a simplification of the actual procedure. In a real system, in addition to other, irrelevant action codes, we would probably have to capture a number of errors before the common sequence became obvious.

The software FIFO was limited to 8 bytes to make this example manageable. Even with a larger FIFO, there would be a temptation when this problem appeared just to increase the size of the FIFO. This could have the effect of making the problem less frequent without fixing it, making it that much harder to find. Worse, a larger FIFO might mask the problem until another code change did something to reduce overall throughput, when the problem could reappear and be blamed on the unrelated changes.

Debugging a Problem in the PC Environment

Figure 9.3 shows the key components of this system. The system could contain from one to three proprietary ISA-based intelligent I/O boards (depending on customer requirements). All the boards shared a single interrupt in the PC-based system. To prevent interrupt lockup (see Chapter 8), the software would test all the slave boards when an interrupt occurred, and service any pending interrupts. The interrupt handler was fairly simple: it set a flag to indicate which DSP had requested the interrupt, checked the other boards (if installed) for pending interrupts, and exited. The application program took care of reading and

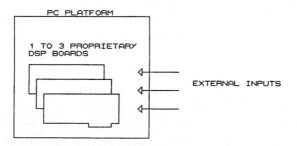

Figure 9.3 Components of PC Platform System

processing the data. To minimize the possibility of a missed interrupt, the interrupt handler sent an EOI to the controller after the first interrupt, so the controller would detect another edge.

The proprietary board accepted signals from an external source and a DSP on the board performed waveform analysis, passing the result to the PC CPU. The majority of systems used only one of the custom boards. Some used two boards, and a very few needed three.

The software had interrupt handlers for all unused interrupts, and also produced an error message for valid interrupts for which there was no interrupt pending. For example, if the common interrupt from the custom boards occurred, but none of the boards was requesting an interrupt, the ISR would produce an error code.

The systems worked in the field, with a few two- and three-DSP users reporting occasional Unexpected Interrupt messages. Since this message did not stop system operation, the problem was given a low priority, in favor of improvements requested by some customers.

In the initial system design, the PC CPU would individually command each DSP to perform a waveform analysis, and the DSP board would interrupt the CPU when processing was done.

After the original programmer left the project, a decision was made to restructure the software in order to meet certain customer requests for performance improvements. The programmer assigned to the task had come from an application-based background; her real-time experience was minimal. The most significant changes made to the software were:

Pipelining of DSP activity; the PC CPU would initiate the DSP to perform a "search" function, where the DSP would monitor the input waveforms for particular patterns and notify the PC CPU when a correlation was detected.

Addition of the "search" function to the DSP firmware.

Functionality was added to the DSP ISRs, making execution time considerably longer.

The new software was tested in the lab with no problems and then sent to several beta sites. The beta sites, all unfortunately using single- or dual-DSP configurations, experienced no problems. The new software was implemented for all sites, and three-DSP customers immediately began experiencing frequent lockups of the system. The PC would continue to operate, but the DSP boards would appear to quit working. The customer who had been the major driver behind the new functionality stopped shipments of new systems until the problem was resolved. Customers who could went back to the previous software version.

Most of the field engineers were trained only to load new software and swap boards to fix problems. At one site, an industrious field engineer with more technical experience was called in to load the previous software version and convinced the customer to let the system run with the new software until it failed again. Measuring test points with an oscilloscope and communicating by telephone with a hardware engineer at the factory, the field engineer determined that the DSP boards were still running but that the common interrupt line was stuck in the high (active) state.

At that point, the software engineer took over and asked the field engineer to examine various memory locations using a debugger program that was included as part of the system software. They were able to determine three things:

- The interrupt was stuck because DSP board 2 was attempting to assert the interrupt.
- The last DSP interrupt serviced by the PC CPU had both DSP 1 and DSP 3 requesting an interrupt, but not DSP 2.
- The pending interrupt was valid; DSP 2 had data in a buffer to pass to the PC CPU.

The software engineer in charge of the PC software insisted that the problem must be due to the DSP firmware change; the hardware engineer in charge of the DSP hardware/firmware insisted that this was impossible. At this point, a "Tiger Team" was formed to determine the cause of the problem. The first order of business was to attempt to reproduce the problem in the engineering lab, where it would be easier to solve. One system was set up to run the normal test suite 24 hours a day, and a second system was dedicated to stress testing in hopes of forcing a failure.

One remote site, run by the previously mentioned major customer, had two of the systems, using one as a backup. They agreed to allow the factory engineers to diagnose the problem on their spare system. After a week of testing, the hardware and software engineers had airline tickets and packed bags for the customer site, when the problem appeared on the stress test unit in the lab. An emergency teleconference was held with the customer, who agreed to allow the problem to be pursued in the lab if it could be reproduced reliably.

It was determined that a particular input sequence caused the problem to occur. A logic analyzer was connected to the common interrupt lines and to the interrupt logic on the individual DSP boards. The analyzer was set to trigger if the common interrupt stayed active for more than a few milliseconds. Figure 9.4 shows the resulting waveforms.

It was determined from this data that the PC ISR was servicing the first DSP interrupt that occurred. The software would then check the other two DSPs and service them if they had requested an interrupt. The problem occurred when the following sequence of events occurred:

DSP 1 asserted an interrupt.

While the software was servicing the interrupt, DSP 3 asserted an interrupt.

After servicing the first interrupt, the ISR checked DSP 2 and DSP 3, in that order, for pending interrupts.

The problem occurred if DSP 2 asserted the interrupt after it was checked by the software, but before the software had serviced the asserted DSP 3 interrupt.

A fix was implemented whereby the software was modified so that every time the ISR serviced a DSP interrupt, it rechecked the other two. In the problem scenario, the ISR would service DSP 1, then DSP 3, then check DSP 1 and DSP 2 again. Finding the missed DSP 2 interrupt asserted, it would service that interrupt, check the other two again, and return.

While looking for this problem, it was noticed that the Unexpected Interrupt problem could occur if a second DSP asserted the interrupt after the EOI had been issued for the first interrupt, but before the CPU checked for pending interrupts from the second DSP. The CPU would poll the second DSP, find the interrupt request

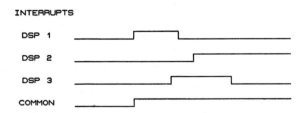

Figure 9.4 Interrupt Lockup Logic Analyzer Waveforms

asserted, and service the interrupt. However, the interrupt controller had already latched the interrupt request, so when interrupts were reenabled at the end of the ISR, the controller would request an interrupt. The CPU, having already cleared the condition that caused the interrupt, would find no interrupt condition pending and would issue the Unexpected Interrupt message.

Three things can be learned from this problem:

1. The original software actually had a window where the stuck interrupt could occur, but the synchronizing action that resulted from having the CPU initiate all operations made the event unlikely to occur. In addition, the simplicity of the original ISR made the time window extremely small.
2. The modified software opened the time window in which the problem could occur by making the ISR more complex. In addition, the free-running nature of the new DSP code made the DSP interrupts less synchronous to the main CPU timing, thereby making the problem more likely to occur.
3. The original suite of tests were intended to verify that the software and hardware worked as intended, but not to stress the system.

Figure 9.5 shows the timing relationships for the original code and the windows in the new code where the problems could occur.

An International Incident

We had sold a large, multiprocessor system in the United States for some time, and the first international user was ramping up their marketing effort. This OEM system contained numerous processors and was controlled by a computer system supplied by the customer. The customer sold complete systems—consisting of the OEM machine, their software, and their computer system—to the end user. Developing and integrating the control software was a lengthy process that had to begin well before product introduction.

We were contacted by a customer in Paris, France, who had encountered intermittent problems with the equipment during their integration testing. They had a schedule to introduce their system based on our machine, and that schedule was in jeopardy due to the machine problems.

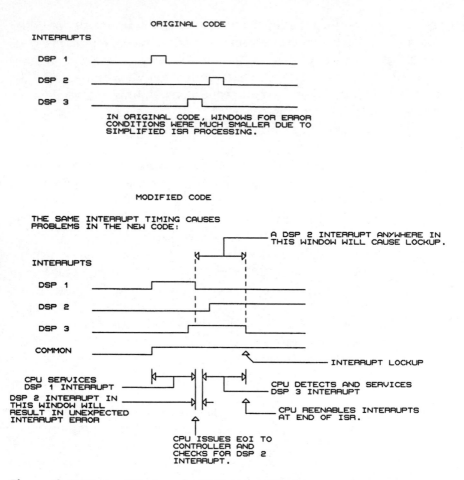

Figure 9.5 Timing Windows for PC Interrupt Problem

An emergency meeting was called where management, engineering, and marketing discussed the problem at length. The following points were made:

- No other customers had seen these problems.
- The factory had not seen these problems.
- The problems must therefore be related to the use of 50 Hz power in France, which is at a higher voltage than U.S. power.
- A Tiger Team must be dispatched to the customer site immediately.

Within a week, a team consisting of myself, a software engineer, and a technical support person from the marketing organization was sent to France. We had already shipped a bulky logic analyzer and a microprocessor emulator ahead so they would have cleared customs by the time we arrived.

We observed that the machines would lock up, displaying a fatal error message, and that the customer's software did not need to be running for the error condition to occur. Most of the symptoms occurred when starting the machine. We tried numerous fixes related to the 50 Hz power, to no avail. The machine drive mechanism was as shown in Figure 9. 6.

To start the machine, the software would set a bit in a register, which turned on a relay, which in turn applied power to the synchronous AC motors. Exhausting our supply of power ideas, we tried disconnecting the motors. We built interlock bypasses so the machine would think the motors were still connected, and found that the problem still occurred. Next, we disconnected the relay, on the assumption that perhaps the flyback voltage was causing problems. The problem still occurred. At this point, the software was driving an open circuit consisting of the control register and a transistor driver for the disconnected relay.

Now certain that the power had nothing to do with the problems, we connected an emulator to the subsystem that was generating most of the errors. After a day of setting breakpoints and tracing program

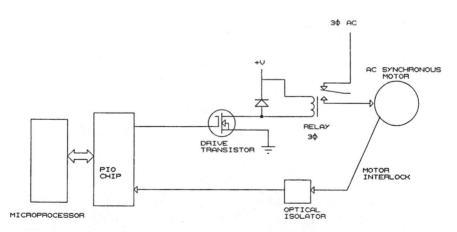

Figure 9.6 Motor Drive Logic

flow, we discovered a race condition in the software. Now that the door was open, we were able to trace some of the other symptoms to software as well. These problems were mostly introduced with a new revision of software, which the French customer had been one of the first to receive.

The introduction of new software explained why other customers had not seen these problems. The remaining question was why the factory had not. Upon our return to the United States, I asked the lead technician in the factory what problems they were experiencing in bringing up the machines. He immediately mentioned the symptoms that we had seen most often in France. The factory had been seeing these symptoms for over a month. Either these symptoms were not being reported to engineering, or someone in engineering management was not listening.

During this trip, we worked six days a week, starting early in the morning and ending late in the afternoon, during one of the coldest winters Paris had experienced in years. One of the managers in the United States chewed us out for changing money at the hotel desk, where the exchange rate was "terrible." I pointed out that we were starting work before the banks opened and returning long after they had closed, and asked if he would prefer that we take a few hours off every couple of days to save a few dollars on the exchange rate. He insisted that we should somehow find a way to exchange money at a lower rate without changing our work hours. I was never sure how we were supposed to do that in a cost-effective manner.

Lessons learned:

- Look before you leap. The trip to Paris cost thousands of dollars in direct travel expenses, plus the engineering time that was lost to other projects while we were away. Had we known that the factory was seeing the same symptoms, the trip could have been avoided and the problems solved without chasing the red herring of 50 Hz power. Besides, we could have saved the few dollars we wasted changing money at the hotel.
- Have communication. Had there been a formal communication path between engineering and manufacturing, the entire incident could have been avoided.
- Don't cling to a bad theory. We wasted nearly a week chasing nonexistent power problems because we were told that other customers and the factory had not seen this problem.

- Fixing the bugs is easier in your own lab than it is in a foreign country where you suffer from jet lag, language problems, and unfamiliar food. Also, you won't waste dollars changing money at the hotel.
- Nothing is impossible for the manager who doesn't have to do it himself.

Christmas in England (Almost)

A site in England had recently begun experiencing intermittent problems in a certain board in their system. Sometimes the boards would malfunction; other times a particular IC would fail, ruining the board. The boards were failing at the rate of a few per week, and the customer was justifiably unhappy about the expense of replacing them, not to mention the down time required to do so.

The system being used had originally been introduced in the United States, and a European version was never built. Consequently, use in England required a motor/generator arrangement to produce the 60 Hz power required by the equipment. At this particular site, which happened to be well south of London, the motor/generator was located in the basement of the building. The customer used several of the machines, connected via a LAN to other equipment. Figure 9.7 shows the connection arrangement.

This machine was designed before I started with the company, and I had never worked on one or looked at the schematics. Nevertheless, I found out about the problem about 11:00 A.M. one day, and by 3:30 P.M. that same day I was on the way to England with a marketing technician who was familiar with the machines in question. To avoid customs delays, the local field engineer for the site had agreed to rent a logic analyzer locally. To become familiar with the equipment, I read the service manual on the flight over.

Within a couple of days, we were able to determine what condition (a noise burst) indicated the presence of the problem, and we could trigger the analyzer when it occurred. This took us no closer to a solution, however.

The day that we determined how to trigger the analyzer, I noticed that the analyzer, which usually triggered every few minutes, did not trigger at all during lunch. I also noticed that it quit triggering at about 4:55 P.M.

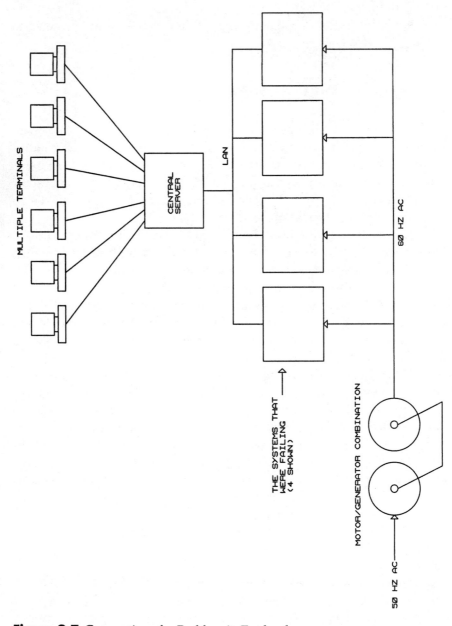

Figure 9.7 Connections for Problem in England

This made me suspect that the problem was in the power and/or grounding of the machine, triggered by a building air-conditioning compressor or other large surge load. Since the machines had so many interconnected grounds, including the LAN connections and the power cabling, it was difficult to determine what the source of the problem was. We had an electrician inspect the motor/generator connections, especially the safety grounds, and he determined that all was as it should be.

We left the logic analyzer connected overnight. When we came in the next morning, it had gone all night without triggering. At about 8:05, it began triggering, and continued to do so every few minutes. Now I was sure that the problem was somewhere in the grounding.

We had discovered that the site cafeteria prepared an excellent breakfast, and had developed the habit of taking advantage of it. One morning, I was watching construction of a new building across the street through the cafeteria windows, when I saw flash of light.

"I think I just found our problem," I said to my companion.

"What?" he asked.

I pointed out the window. "Arc welders," I replied. "The current is getting into our grounds somehow."

It was two days before Christmas and the site, as well as our hotel, was closing for the holidays. Not wanting to sleep in the streets or to spend Christmas away from our families, we returned to the United States. I was told later that this disappointed at least one manager, who thought us disloyal for not staying over the holidays. He wasn't volunteering to relieve us, of course.

After Christmas, another engineer went to the site, came to the same conclusion we did, and temporarily ran the safety ground out a window to another location. The problems went away. Since the machines had been sold with the provision that the customer provide clean power to operate them, the ball was now in the customer's court. Although they had originally been adamant about having the problem fixed, they now decided to live with the expense of replacing the failed boards until the new building was completed, instead of isolating and fixing the problem. When the welding on the new building stopped, so did the failures.

Although this problem did not result in an upset of the microprocessor, it could have been the processor as easily as the logic, depending on how the ground paths had been designed.

Lessons learned:

- Don't ignore symptoms. During this trip, we were in daily contact with marketing and engineering management in the United

States. When I announced the discovery that the problem disappeared every day at 5:00 and reappeared at 8:00 the next morning, it was met with skepticism to the point of derision. Nevertheless, this was a key fact in solving the puzzle.

- Sometimes you have to ignore the skeptics. When you have facts versus facts or opinions versus opinions, either side can be right. When you have facts versus opinions, as in the 8-to-5 nature of the problem, go with the facts.

- If your equipment must operate in a harsh environment, especially one that is electrically noisy, pay careful attention to shielding and grounding. Avoid ground loops, especially in low-level analog circuits. Add a watchdog timer to the processor so it recovers from upsetting transients.

California Dreamin'

A site in Los Angeles was experiencing intermittent problems with a large machine. This particular machine had originally been designed with discrete logic, and microprocessor control was added to portions of the machine when a later model came out. The line was replaced with a completely redesigned and microprocessor-controlled version soon thereafter, so not many of the hybrid machines were ever sold.

The system at this site had a number of problems, and the company ended up sending five engineers, including myself, to work on it. In addition to the engineers, there were two people from the marketing group at corporate headquarters, the site representative from marketing who had sold the account, and a local/regional marketing manager who came by once a day or so to see how things were going. The site rep appeared to me to be spending a good deal of his time trying to convince the customer that he understood and was directing the technical work that was being done, although in one instance, he was overheard referring to "pull-ups" as "jump-ups."

The site was located in a section of Los Angeles that wasn't where I'd normally want to spend a lot of time. It wasn't all that bad during the day, but you didn't want to wander around the streets at night. The engineers were all staying at a local hotel near the site. The marketing people stayed somewhere in Beverly Hills.

We found a couple of problems in the parts of the machine that were controlled by logic, and finally got down to the processor part. The processor was an 8031, and we did not have an emulator for it.

Although we did not find it out until later, the problem was in an interrupt input to the 8031. This particular interrupt was programmed as edge-sensitive because one of the interrupt inputs originated from an optical interrupter and could not be cleared by software.

The interrupt was shared, open-collector, between two different devices. Figure 9.8 shows how the timing was supposed to work. As shown in the figure, the pulse interrupt set a flipflop, the output of which connected to a port pin on the 8031. When the ISR executed, it read the flipflop to see which interrupt was set. Since the second interrupt (interrupt 2 in the figure) resulted from an event that was initiated by the optical interrupter pulse, there was no chance of the interrupts overlapping. If the flipflop was set, interrupt 1 (the optical interrupter) was the source, and if the flipflop was reset, the other interrupt was assumed to be the source. The CPU would reset the flipflop after reading it. As shown in the figure, if the source of the interrupt was interrupt 2, the CPU would reset the condition that caused the interrupt when finished.

The interrupts were used to initiate certain machine operations. Interrupt 1 was used to tell the 8031 to image a previously loaded ID number in a logic-controlled subsystem. Interrupt 2 told the 8031 that a new ID number could be loaded. The effect was that a number was loaded on interrupt 2 and then imaged on the following interrupt 1. In this particular case, the 8031 would be told a starting number and would increment the number for each item processed, until given a new starting number.

The symptom we were seeing was that two successive items would have identical ID numbers, and the ID numbers for all subsequent items would be off by one. It was the worst kind of intermittent, where we could not see the symptom until hours after the problem occurred, when the results of the entire run were available.

A brainstorming session resulted in some possible scenarios:

- Interrupt 2 was not being generated correctly.
- Interrupt 1 latency was too long.
- The 8031 somehow failed to increment the number
- The possibility that the other subsystem was not imaging the number was discarded, as this would cause only one number to be incorrect and would not affect subsequent numbers.

We determined that we needed to know what was happening around the problem. The only thing we knew for certain was that the

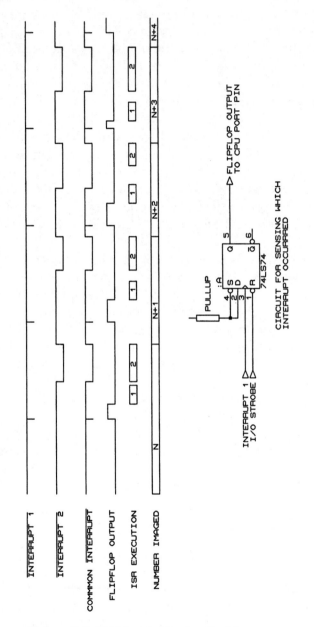

Figure 9.8 8031 Latch/Timing Problem: Expected Timing

problem always started with two identical ID numbers. We went to a local electronics store and purchased parts to build a test circuit, which was wired into the system. The circuit consisted of a latch and an 8-bit comparator. The latch captured the least significant byte of the previous ID number. The comparator (with some synchronizing logic) compared the latch output to the least significant byte of the current ID number. A trigger was generated for the logic analyzer if the two numbers were ever the same.

When the analyzer triggered, we got a trace that looked something like that of Figure 9.9. The problem was that the ISR for interrupt 1 took varying amounts of time to begin execution, depending on the state of the machine when the interrupt occurred. Very rarely, the ISR would respond to interrupt 1 after interrupt 2 had already occurred. The 8031 interrupt logic clears the request flag when the interrupt is serviced, so if interrupt 2 occurred during the servicing of interrupt 1, the second interrupt would still be recognized. However, there is only one edge detection stage, so if interrupt 2 occurred before interrupt 1 was serviced, the second interrupt was lost, resulting in interrupt 2 being missed. The reason the interrupt didn't lock up with interrupt 2 always active was that a quirk in the logic for interrupt 2 cleared it when the optical interrupter pulse occurred, before interrupt 1 was generated (see timing detail in Figure 9.9).

Once the problem was identified, a look at the code revealed how the machine state caused the error. Unfortunately, it was not a coding bug, but just a stack-up of high-priority conditions that occasionally occurred in another ISR.

Various solutions were proposed, from completely rewriting the code right there on site (impractical) to changing the hardware so the CPU could momentarily gate the interrupt off (no spare port pins; extensive hardware changes). It was finally suggested that, since the error condition caused the 8031 to just barely miss the interrupt, the code should be modified to test the interrupt input after servicing interrupt 1, to see if it was still active. If so, go ahead and service interrupt 2. This was possible because the 8031 interrupt inputs double as port pins, and can be directly tested for a one or a zero. This was a simple solution that solved the problem.

Soon after we returned, a memo was circulated from corporate headquarters congratulating the marketing salesman for solving the problems at the site. No mention was made of the contribution from engineering, although local management assured us that they knew who really did the work. That year, the salesman went on the company's

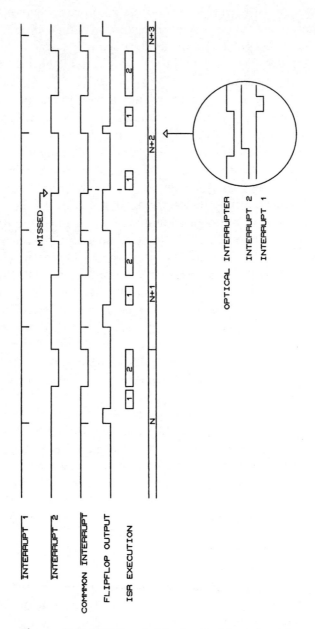

Figure 9.9 8031 Latch/Timing Problem: Actual Timing

annual Hawaii trip. Number of engineers who went to the site and also went to Hawaii: zero. What did you expect?

Lessons learned:

- If you can't avoid traveling with marketing, insist that you at least get to stay in the same hotel they do.
- The marketing people will always figure a way to get all the credit.
- Once in a while, even with sophisticated debugging aids, you still have to build a tool to see what's really going on.
- Sometimes the simplest solutions are the best.

Would You Like ICE in Your Drink?

In most cases, and for development projects of any size, having the right tools is cheaper than the savings obtained by working with the cheapest tools. A couple of days spent chasing a problem that can be solved in half a day with an ICE usually makes the emulator cheap by comparison. Forcing developers to work with the stare-at-the-code method of debug, although it makes the capital equipment budget look better, is usually very shortsighted and extremely unproductive from a business point of view.

For this reason, people sometimes question the wisdom of using simple tools such as action codes when much more powerful tools are available. My answer comes from my experience, much of which has involved working with systems where an emulator could not be used some or all of the time.

A case in point is a robotic arm for which I developed a component. The arm moved a suction probe in and out of various liquids. The part I worked on was a sensor, based on one of the Microchip PIC devices. The circuit board was mounted on the arm and moved with it.

I elected not to use an emulator during development, for three reasons: first, the design was simple enough that I felt an emulator was unnecessary. Second, I thought the advantages of using the PIC outweighed the fact that purchase of an emulator would not have been approved. Third, the clearances around the board were very small. At one point, the board had to move up through a fairly narrow slot in the frame, which would have sheared the emulator off and left it in the drink, so to speak.

I used the code in Listing 9.1 to output a self-clocking data stream on bit RB7 of the PIC device. The output format consisted of a positive clock pulse for each bit, followed by a low time equal to the clock pulse time, then another positive pulse if the data bit was a '1'. Readers familiar with disk drive formats will recognize this as the original FM encoding method used on early disks. The decoding logic, embedded in an FPGA, looked for a clock pulse, then started a counter that timed out in the middle of the data pulse. The counter timeout clocked the data pulse or lack thereof into a shift register.

During development of the code, which took place on the bench, the diagnostic was used to output values that indicated what the program was doing. Once the program was working, certain design parameters had to be set in actual operation. Therefore, on the operating arm, the diagnostic consisted of an 8-bit A/D value and 4 status bits from which everything else in the system could be inferred. The diagnostic output and a ground were run out to the decoder board, and the data was captured on a logic analyzer.

Lessons learned:

> Although having the right tools is important, sometimes you just can't. There are also times when a simple microprocessor is the best solution to a problem, even if the best tools aren't available.
>
> ...But don't let anyone use this argument to talk you into working hours of overtime so they can save a few dollars from the equipment budget by not purchasing the right tools for a major project.

Listing 9.1 PIC Synchronous Serial Diagnostic Output Code

```
;------------------------------------------------------------
; Synchronous serial diagnostic output.
; Sends 16-bit diagnostic word in DIAGLSB/DIAGMSB,
; to RB7.  MSB of DIAGMSB is sent first, LSB of DIAGLSB
; is sent last.  DIAGBITS is a bit counter variable.
; Output is clock bit followed by data clock if bit
; was a '1', no data clock if bit was a '0'.  Works
; like FM disk data:
;
;           _   _
;    1: __/ \__/ \__
;
;           _
;    0: __/ _____
;
; This makes the data self-clocking.
; Data is transmitted in the order:
;   DIAGLSB D0 --> DIAGMSB D7.
;------------------------------------------------------------
```

Listing 9.1 PIC Synchronous Serial Diagnostic Output Code *(continued)*

```
diagnostic:

        movlw 8             ; bit count
        movwf diagbits      ; store at bit counter

        ; Transmit LS byte first
diaglsblp:
        bsf portb, 7
        rrf diaglsb,f       ; rotate DIAGLSB.0 into cy
        bcf portb, 7        ; clock bit falling edge
        btfsc status,c      ; check carry
        bsf portb,7         ; set data clock if bit = 1
        nop
        bcf portb, 7         ; clear data clock, set or not.
        decfsz diagbits,f   ; done?
        goto diaglsblp

        ; Transmit MS byte
        movlw 8             ; bit count
        movwf diagbits      ; store at bit counter
diagmsblp:
        bsf portb, 7        ; clock bit rising edge
        rrf diagmsb,f       ; rotate DIAGMSB.0 into cy
        bcf portb, 7        ; clock bit falling edge
        btfsc status,c      ; check carry
        bsf portb,7         ; set data clock if bit = 1
        nop
        bcf portb, 7         ; clear data clock, set or not.
        decfsz diagbits,f   ; done?
        goto diagmsblp

        clrf diagmsb
        clrf diaglsb        ; clear both diagnostic words

        return

;------------------------------------------------------------
;   End of diagnostic output code
;------------------------------------------------------------
```

Appendix 1 _____
Programmer Test Routines

Test 1

Initialize internal 80188 registers, then loop, reading RAM location 0000.

```
; I/O port addresses

uart equ 0 ; uart base addr
uartdata equ uart + 0 ; uart rx/tx data reg
uartie equ uart + 1    ; uart interrupt enable
uartii equ uart + 2    ; uart interrupt id reg
uartctl equ uart + 2   ; uart fifo control
uartlc equ uart + 3    ; line control
uartmc equ uart + 4    ; modem control
uartls equ uart + 5    ; line status
uartdlo equ uart + 0   ; divisor latch - lo - if DLAB = 1.
uartdhi equ uart + 1   ; divisor latch - hi - if DLAB = 1.

DIAGNOST equ 312 ; diagnostic output strobe address

; 80188 peripheral control block reg addresses
pcb equ 0ff00h
t2cnt equ pcb + 60h
t2con equ pcb + 66h
t2cmp equ pcb + 62h
t1cnt equ pcb + 58h
t1cmpa equ pcb + 5Ah
t1con equ pcb + 5Eh
PACS equ pcb + 0A4h ·
MPCS equ pcb + 0A8h
lmcs equ pcb + 0A2h
umcs equ pcb + 0A0h

    org 0
```

```
; Initialize segment registers
INIT: mov ax,0
   mov ds,ax      ; data RAM area
   mov ss,ax      ; stack seg
   mov ax,1000h
   mov es,ax      ; program data area
   mov sp,0FFFFh  ; top of stack = 0:FFFF.

; Hardware initialization:
; Set LCS to 256k (LMCS @ FFA2h = 3FF8h)
; Set PCS block to start at 0000, with 1 wait state
; PACS @ FFA4h = 0038h
; MPCS @ FFA8h = 81B8h

   mov dx,lmcs
   mov ax,3FF8h
   out dx,al
   mov dx,pacs
   mov ax,0038h
   out dx,al
   mov dx,mpcs
   mov ax,81B8h
   out dx,al

; Set UART up for Rx clock = tx clock, 19200 baud, n81.
; UART input clock is processor CLKOUT, or 7.373 Mhz.
; Divisor is 7.373 Mhz / (19200 x 16), or 24.
; UARTLC = 83h
; UARTDLO = 23d
; UARTDHI = 0
; UARTLC = 3
; Enable UART receive data interrupt
; UARTIE = 01
; Initialize other UART registers
; UARTMC = 2

   mov dx,uartlc
   mov al,83h
   out dx,al
   mov dx,uartdlo
   mov al,24
   out dx,al
   mov dx,uartdhi
   mov al,0
   out dx,al
   mov dx,uartlc
   mov al,3
   out dx,al
```

```
    mov dx,uartie
    mov al,1
    out dx,al
    mov dx,uartmc
    mov al,2
    out dx,al

; At this point, the UART baud clock output should be
; 307200 Hz (3.2 microseconds). The following code
; will just loop, reading RAM location 0, so the
; RAM CS signal can be verified on a scope.

loop1: mov si,00
    mov ax,[si]
    mov dx,diagnost
    out dx,al
    jmp loop1

; Reset Vector/High Memory Initialization
; Set UCS to 64k, no waits
; UMCS @ FFA0h = F038h
; Jump to init code.

org 1FF0h      ; init at top of ROM.

    mov dx,umcs
    mov ax,0F038h
    out dx,al
    jmp init       ; long jump to FE00:0000
```

Test 2

Verifies 128k SRAM

```
; Serial programmer TEST2 - 128k SRAM test.
; Just writes incrementing values to all 128k
; of RAM, then verifies, then changes base
; and repeats.

; I/O port addresses

CTLREG equ 272 ; ACT374 control reg
      ; D0 D1 LED
      ; 0 0 off
      ; 0 1 red
      ; 1 0 green
      ; 1 1 off
```

```
        ; D2: ROM -OE
        ; D3: ROM -CE
        ; D4-D7: CTL 1 - CTL 3

DIAGNOST equ 312 ; diagnostic output strobe

; 80188 peripheral control block reg addresses.
pcb equ 0ff00h
t2cnt equ pcb + 60h
t2con equ pcb + 66h
t2cmp equ pcb + 62h
t1cnt equ pcb + 58h
t1cmpa equ pcb + 5Ah
t1con equ pcb + 5Eh
PACS equ pcb + 0A4h
MPCS equ pcb + 0A8h
lmcs equ pcb + 0A2h
umcs equ pcb + 0A0h


org 0

; Initialize segment registers
INIT: mov ax,0
      mov ds,ax       ; data RAM area
      mov ss,ax       ; stack seg
      mov ax,1000h
      mov es,ax       ; program data area
      mov sp,0FFFFh ; top of stack = 0:FFFF.

; Hardware initialization:
; Set LCS to 256k (LMCS @ FFA2h = 3FF8h)
; Set PCS block to start at 0000, with 1 wait state
; PACS @ FFA4h = 0038h
; MPCS @ FFA8h = 81B8h

      mov dx,lmcs
      mov ax,3FF8h ; 1 internal wait state
      out dx,al
      mov dx,pacs
      mov ax,0038h
      out dx,al
      mov dx,mpcs
      mov ax,81B8h
      out dx,al

; RAM test. Logic:
; Initialize seed to 0
; loop1:
```

```
;   Initialize pointer to 0000
;   Initialize segment to 0000
;   Move seed to value
;   loop2:
;    Write value to segment:pointer
;    Increment value
;    Increment pointer
;   loop2 until pointer = 0
;  End of loop2
;  If segment = 0000,
;   Set segment to 1000h
;   Repeat loop1.
; (Now verify the data)
;  Initialize segment:pointer to 0:0
;  Move seed to value
;  loop3:
;   Read RAM at segment:pointer
;   if result <> value, hang in loop reading error
;   Increment value
;   Increment pointer
;   loop3 until pointer = 0
;  End of loop3
;   If segment = 0,
;    set segment = 1000h
;    Repeat loop3
;  Increment seed
;  Loop1.
;
;Register usage:
;  DI = seed
;  SI = pointer
;  ES = segment register
;  bx = value
;  ax = working register

        mov di,0
loop1:
        mov si,0
        mov ax,0
        mov es,ax
        mov bx,di
        mov dx,ctlreg
        mov al,0FDh
        out dx,al ; toggle LED color
loop2:
        mov es:[si],bx
        inc bx
        inc si
```

```
        inc si
        jnz loop2
        mov ax,es
        cmp ax,0      ; segment = 0?
        jnz verify    ; no, must be 1000, time to verify.
        mov ax,1000h  ; yes, set seg to 1000 and repeat.
        mov es,ax
        jmp loop2
verify:
        mov dx,ctlreg
        mov al,0FEh
        out dx,al ; toggle LED color
        mov bx,di
        mov si,0
        mov ax,0
        mov es,ax
loop3:
        cmp bx,es:[si]
        jnz error
        inc bx
        inc si
        inc si
        jnz loop3
        mov ax,es
        cmp ax,0        ; segment = 0?
        jnz tstdone     ; no, must be 1000, pass done
        mov ax,1000h    ; yes, set seg to 1000 and repeat
        mov es,ax
        jmp loop3
tstdone:                ; pass done, incr seed and repeat
        inc di          ; incr seed
        mov dx,diagnost
        out dx,al       ; trigger diagnostic port
        jmp loop1       ; repeat entire test with new seed value

error:
        mov ax,es:[si] ; just keep reading the bad location
        jmp error

; Reset Vector/High Memory Initialization
; Set UCS to 64k, no waits
; UMCS @ FFA0h = F038h
; Jump to init code

org 1FF0h     ; init at top of ROM

    mov dx,umcs
    mov ax,0F038h
    out dx,al
    jmp init      ; long jump to FE00:0000
```

TxTest1

Code to test serial transmitter hardware on the serial programmer. Just transmits character "A" continuously.

```
; EQUATES
; I/O port addresses

uart equ 0 ; uart base addr.
uartdata equ uart + 0 ; uart rx/tx data reg
uartie equ uart + 1   ; uart interrupt enable
uartii equ uart + 2   ; uart interrupt id reg
uartctl equ uart + 2  ; uart fifo control
uartlc equ uart + 3   ; line control
uartmc equ uart + 4   ; modem control
uartls equ uart + 5   ; line status
uartdlo equ uart + 0  ; divisor latch - lo - if DLAB = 1.
uartdhi equ uart + 1  ; divisor latch - hi - if DLAB = 1.

CTLREG equ 272 ; ACT374 control reg
       ; D0 D1 LED
       ; 0 0 off
       ; 0 1 red
       ; 1 0 green
       ; 1 1 off
       ; D2: ROM -OE
       ; D3: ROM -CE
       ; D4-D7: CTL 1 - CTL 3

DIAGNOST equ 312 ; diagnostic output strobe

; 80188 peripheral control block reg addresses.
pcb equ 0ff00h
t2cnt equ pcb + 60h
t2con equ pcb + 66h
t2cmp equ pcb + 62h
t1cnt equ pcb + 58h
t1cmpa equ pcb + 5Ah
t1con equ pcb + 5Eh
PACS equ pcb + 0A4h
MPCS equ pcb + 0A8h
lmcs equ pcb + 0A2h
umcs equ pcb + 0A0h

; Variable Definition
RAM equ 0

    org 0

; Initialize segment registers
```

```
INIT: mov ax,0
     mov ds,ax      ; data RAM area
     mov ss,ax      ; stack seg
     mov ax,1000h
     mov es,ax      ; program data area
     mov sp,0FFFFh  ; top of stack = 0:FFFF.

; Hardware initialization:
; Set LCS to 256k (LMCS @ FFA2h = 3FF8h)
; Set PCS block to start at 0000, with 1 wait state
; PACS @ FFA4h = 0038h
; MPCS @ FFA8h = 81B8h

     mov dx,lmcs
     mov ax,3FF8h
     out dx,al
     mov dx,pacs
     mov ax,0038h
     out dx,al
     mov dx,mpcs
     mov ax,81B8h
     out dx,al

; Set UART up for Rx clock = tx clock, 19200 baud, n81.
; UART input clock is processor CLKOUT, or 7.373 Mhz.
; Divisor is 7.373 Mhz / (19200 x 16), or 24.
;  UARTLC = 83h
;  UARTDLO = 24d
;  UARTDHI = 0
;  UARTLC = 3
; Enable UART receive data interrupt
;  UARTIE = 01
; Initialize other UART registers
;  UARTMC = 2

     mov dx,uartlc
     mov al,83h
     out dx,al
     mov dx,uartdlo
     mov al,24
     out dx,al
     mov dx,uartdhi
     mov al,0
     out dx,al
     mov dx,uartlc
     mov al,3
     out dx,al
     mov dx,uartie
```

```
        mov al,1
        out dx,al
        mov dx,uartmc
        mov al,2
        out dx,al

; At this point, the UART baud clock output should be
; 307200 Hz (3.2 microseconds).

; (Transmit service)
; If UART Tx register empty (UARTLS bit 5 = 1)
;   If CTS active (UARTMS bit 4 = 1)
;     If TxFIFO not empty (TWPOINT <> TRPOINT)
;       Write [TRPOINT] to UART tx register
;       Incr TRPOINT, wrap to beginning of buffer.

loop1:
        mov dx,UARTLS
        in al,dx
        and al,020h      ; TxREADY okay?
        jz loop1         ; no, loop
        mov ax,'A'       ; yes, char to xmit
        mov dx,UARTDATA
        out dx,al        ; transmit a byte from the FIFO.
        mov dx,diagnost
        out dx,al
        jmp loop1

; Reset Vector/High Memory Initialization
; Set UCS to 64k, no waits
; UMCS @ FFA0h = F038h
; Jump to init code.

        org 1FF0h ; init at top of ROM.

reset:
        mov dx,umcs
        mov ax,0F038h
        out dx,al
        jmp init ; long jump to FE00:0000
```

TxTest2

Tests RS-232 Tx and Rx hardware on the serial programmer. Transmits
and checks each char, 0-256. Requires Tx be wrapped to Rx at the serial
I/O connector. Toggles LED while testing, leaves LED red if error.

```
; EQUATES

; I/O port addresses

uart equ 0              ; uart base addr.
uartdata equ uart + 0 ; uart rx/tx data reg
uartie equ uart + 1    ; uart interrupt enable
uartii equ uart + 2    ; uart interrupt id reg
uartctl equ uart + 2   ; uart fifo control
uartlc equ uart + 3    ; line control
uartmc equ uart + 4    ; modem control
uartls equ uart + 5    ; line status
uartdlo equ uart + 0   ; divisor latch - lo - if DLAB = 1.
uartdhi equ uart + 1   ; divisor latch - hi - if DLAB = 1.

CTLREG equ 272 ; ACT374 control reg
     ; D0 D1 LED
     ; 0 0 off
     ; 0 1 red
     ; 1 0 green
     ; 1 1 off
     ; D2: ROM -OE
     ; D3: ROM -CE
     ; D4-D7: CTL 1 - CTL 3
DIAGNOST equ 312 ; diagnostic output strobe

; 80188 peripheral control block reg addresses
pcb equ 0ff00h
t2cnt equ pcb + 60h
t2con equ pcb + 66h
t2cmp equ pcb + 62h
t1cnt equ pcb + 58h
t1cmpa equ pcb + 5Ah
t1con equ pcb + 5Eh
PACS equ pcb + 0A4h
MPCS equ pcb + 0A8h
lmcs equ pcb + 0A2h
umcs equ pcb + 0A0h

org 0

; Initialize segment registers
INIT: mov ax,0
     mov ds,ax       ; data RAM area
     mov ss,ax       ; stack seg
     mov ax,1000h
     mov es,ax       ; program data area
     mov sp,0FFFFh ; top of stack = 0:FFFF
```

```
; Hardware initialization:
; Set LCS to 256k (LMCS @ FFA2h = 3FF8h)
; Set PCS block to start at 0000, with 1 wait state
; PACS @ FFA4h = 0038h
; MPCS @ FFA8h = 81B8h

        mov dx,lmcs
        mov ax,3FF8h
        out dx,al
        mov dx,pacs
        mov ax,0038h
        out dx,al
        mov dx,mpcs
        mov ax,81B8h
        out dx,al

; Set UART up for Rx clock = tx clock, 19200 baud, n81.
; UART input clock is processor CLKOUT, or 7.373 Mhz.
; Divisor is 7.373 Mhz / (19200 x 16), or 24.
;   UARTLC = 83h
;   UARTDLO = 24d
;   UARTDHI = 0
;   UARTLC = 3
; Enable UART receive data interrupt
;   UARTIE = 01
; Initialize other UART registers
;   UARTMC = 2

        mov dx,uartlc
        mov al,83h
        out dx,al
        mov dx,uartdlo
        mov al,24
        out dx,al
        mov dx,uartdhi
        mov al,0
        out dx,al
        mov dx,uartlc
        mov al,3
        out dx,al
        mov dx,uartie
        mov al,1
        out dx,al
        mov dx,uartmc
        mov al,2
        out dx,al

; At this point, the UART baud clock output should be
```

```
; 307200 Hz (3.2 microseconds).

; Tx wraparound test: LED is amber while running,
; green if rx char never received, red if received
; char did not verify.
; Does not use interrupts, polls everything.

    mov bl,0 ; beginning transmit value
loop1:
    mov dx,UARTLS
    in al,dx
    and al,020h     ; TxREADY okay?
    jz loop1        ; no, loop
    mov al,bl       ; yes, char to xmit
    mov dx,UARTDATA
    out dx,al       ; transmit a byte
    mov dx,diagnost
    out dx,al
    mov al,0feh     ; make LED red
    mov dx,ctlreg
    out dx,al

; Now wait for the character to receive,
; and check it.

loop2:
    mov dx,uartls
    in al,dx
    and al,01       ; Data ready is UARTLSR D0
    jz loop2

    mov al,0fdh     ; char detected,
    mov dx,ctlreg   ; make LED red, in case char
    out dx,al       ; doesn't verify.

    mov dx,uartdata
    in al,dx
    cmp al,bl
    jnz error
    inc bl
    jmp loop1

error:
    mov dx,diagnost
    in al,dx
    jmp error

; Reset Vector/High Memory Initialization
```

```
; Set UCS to 64k, no waits
; UMCS @ FFAOh = F038h
; Jump to init code.

    org 1FFOh ; init at top of ROM.
reset:
    mov dx,umcs
    mov ax,0F038h
    out dx,al
    jmp init ; long jump to FE00:0000
```

PIO (8255) Test

Writes incrementing data values to 8255 ports A and B. Allows scope
to be used to verify operation.

```
; EQUATES

; I/O port addresses

; PPI is the 8255 which controls the ROM data/ addr lines.
PPI equ 128        ; base address, PCS1
PPIAL = PPI | 0    ; Port A = low addr byte
PPIAH = PPI + 1    ; Port B = hi addr byte
PPIDA = PPI + 2    ; Port C = data I/O
PPICTL = PPI + 3   ; 8255 control reg

DAC0 equ 256       ; DAC0 data reg
DAC1 equ 257       ; DAC1 data reg
DAC2 equ 258       ; DAC2 data reg
DAC3 equ 259       ; DAC3 data reg
DACLATCH equ 264   ; latch all DACs
CTLREG equ 272     ; ACT374 control reg
        ; D0 D1 LED
        ; 0 0 off
        ; 0 1 red
        ; 1 0 green
        ; 1 1 off
        ; D2: ROM -OE
        ; D3: ROM -CE
        ; D4-D7: CTL 1 - CTL 3
DIAGNOST equ 312 ; diagnostic output strobe

; 80188 peripheral control block reg addresses.
pcb equ 0ff00h
t2cnt equ pcb + 60h
```

```
t2con equ pcb + 66h
t2cmp equ pcb + 62h
t1cnt equ pcb + 58h
t1cmpa equ pcb + 5Ah
t1con equ pcb + 5Eh
PACS equ pcb + 0A4h
MPCS equ pcb + 0A8h
lmcs equ pcb + 0A2h
umcs equ pcb + 0A0h

org 0

; Initialize segment registers
INIT: mov ax,0
    mov ds,ax      ; data RAM area
    mov ss,ax      ; stack seg
    mov ax,1000h
    mov es,ax      ; program data area
    mov sp,0FFFFh ; top of stack = 0:FFFF.

; Hardware initialization:
; Set LCS to 256k (LMCS @ FFA2h = 3FF8h)
; Set PCS block to start at 0000, with 1 wait state
; PACS @ FFA4h = 0038h
; MPCS @ FFA8h = 81B8h

    mov dx,lmcs
    mov ax,3FF8h ; 1 internal wait state
    out dx,al
    mov dx,pacs
    mov ax,0038h
    out dx,al
    mov dx,mpcs
    mov ax,81B8h
    out dx,al

; Initialize 8255 for output, all 3 ports.

    mov dx,ppictl ; control reg
    mov al,10000000b ; set all 3 ports as output.
    out dx,al
    mov bx,0 ; 16-bit value to be output
loop1:
    mov al,bl ; write test value to ports a,b
    mov dx,ppial
    out dx,al
    mov al,bh
    mov dx,ppiah
```

```
        out dx,al
        inc bx
        jmp loop1

; Reset Vector/High Memory Initialization
; Set UCS to 64k, no waits
; UMCS @ FFA0h = F038h
; Jump to init code.

        org 1FF0h ; init at top of ROM.

        mov dx,umcs
        mov ax,0F038h
        out dx,al
        jmp init ; long jump to FE00:0000
```

DAC Test

Steps through all 256 possible values at about .1 ms rate. LED toggles
when timeout occurs.

```
; EQUATES
; I/O port addresses

DAC0 equ 256       ; DAC0 data reg
DAC1 equ 257       ; DAC1 data reg
DAC2 equ 258       ; DAC2 data reg
DAC3 equ 259       ; DAC3 data reg
DACLATCH equ 264 ; latch all DACs
CTLREG equ 272     ; ACT374 control reg
        ; D0 D1 LED
        ; 0 0 off
        ; 0 1 red
        ; 1 0 green
        ; 1 1 off
        ; D2: ROM -OE
        ; D3: ROM -CE
        ; D4-D7: CTL 1 - CTL 3
DIAGNOST equ 312 ; diagnostic output strobe

; 80188 peripheral control block reg addresses.
pcb equ 0ff00h
t2cnt equ pcb + 60h
t2con equ pcb + 66h
t2cmp equ pcb + 62h
t1cnt equ pcb + 58h
t1cmpa equ pcb + 5Ah
```

```
t1con equ pcb + 5Eh
PACS equ pcb + 0A4h
MPCS equ pcb + 0A8h
lmcs equ pcb + 0A2h
umcs equ pcb + 0A0h

    org 0

; Initialize segment registers
INIT: mov ax,0
      mov ds,ax     ; data RAM area
      mov ss,ax     ; stack seg
      mov ax,1000h
      mov es,ax     ; program data area
      mov sp,0FFFFh ; top of stack = 0:FFFF.

; Hardware initialization:
; Set LCS to 256k (LMCS @ FFA2h = 3FF8h)
; Set PCS block to start at 0000, with 1 wait state
; PACS @ FFA4h = 0038h
; MPCS @ FFA8h = 81B8h

      mov dx,lmcs
      mov ax,3FF8h
      out dx,al
      mov dx,pacs
      mov ax,0038h
      out dx,al
      mov dx,mpcs
      mov ax,81B8h
      out dx,al

      mov bl,0 ; DAC initial value

loop1:
      mov ax,184 ; .1 ms with 7.37 mhz input
      mov dx,t2cmp
      out dx,al
      mov ax,0
      mov dx,t2cnt
      out dx,al
      mov ax,0C000h ; enable counter
      mov dx,t2con
      out dx,al
loop2:
      mov dx,t2con
      in al,dx
      and al,0020h ; check mc bit (D5)
      jz loop2
```

```
        mov al,bl
        mov dx,DAC0
        out dx,al
        mov dx,DAC1
        out dx,al
        mov dx,DACLATCH ; Move both DAC values from the input
        out dx,al        ; reg to the DAC reg.
        inc bl

        mov dx,diagnost
        out dx,al
        jmp loop1

; Reset Vector/High Memory Initialization
; Set UCS to 64k, no waits
; UMCS @ FFA0h = F038h
; Jump to init code.

        org 1FF0h ; init at top of ROM.
reset:
        mov dx,umcs
        mov ax,0F038h
        out dx,al
        jmp init ; long jump to FE00:0000
```

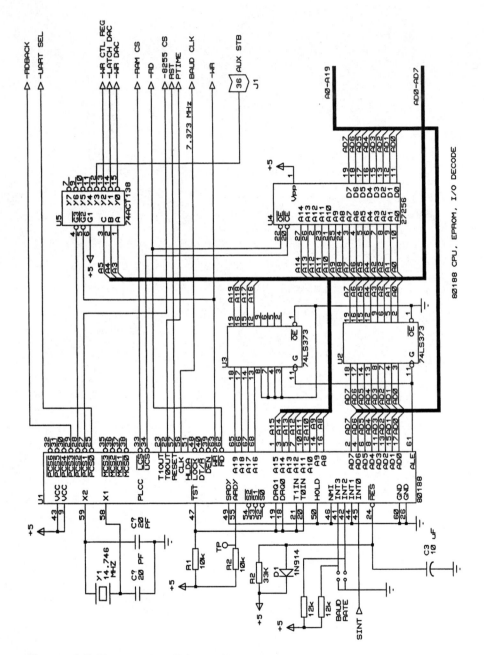

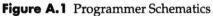

Figure A.1 Programmer Schematics

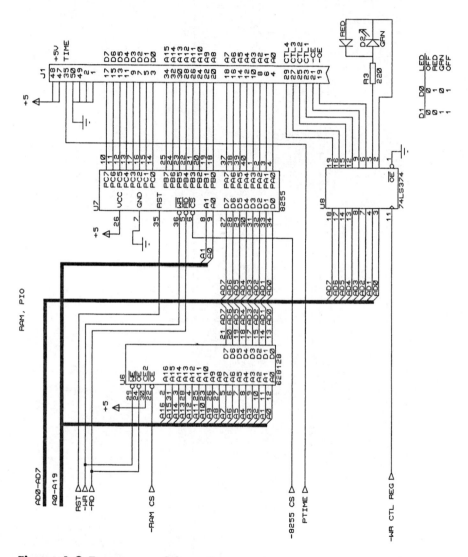

Figure A.2 Programmer Schematics

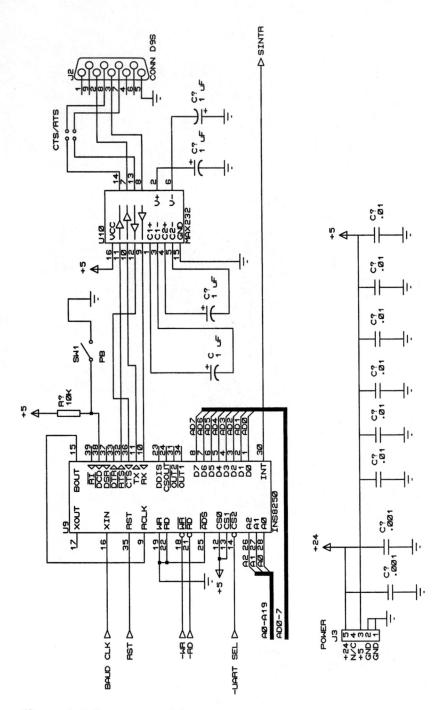

Figure A.3 Programmer Schematics

Glossary

A/D (or A-D): Analog-to-digital converter. An integrated circuit (IC) or subsystem that translates a voltage to a digital word.

Assembler: A language that directly describes machine instructions such as move data to a register, jump to an address, add two registers, and so on. Each microprocessor has a unique machine language, and therefore a unique assembler language.

Blocked: In an application using a real-time kernel, a task is said to be blocked if it cannot run because it is awaiting some event. The event could be a particular interrupt, a specific time delay, or availability of a particular resource.

Breakpoint: A condition that causes the central processing unit (CPU) to execute a software interrupt when a particular address is executed or when a specific event occurs, such as a memory read or write. A breakpoint may be created by substituting an instruction opcode in memory with the software interrupt instruction, or the breakpoint may be implemented in hardware on an emulator.

Context: The state of the processor. In practical terms, this is usually the contents of the processor's registers and condition flags. The processor context is typically saved when it must be restored after a piece of code, such as an interrupt service routine (ISR), finishes executing.

CPU: Central processing unit. Technically the computing core of a microprocessor, the term is commonly used to refer to the microprocessor itself.

D/A (or D-A): Digital-to-analog converter. An IC or subsystem that translates a digital word to a voltage.

Daisy-Chained Interrupts: An interrupt prioritizing scheme where the priority of each peripheral is determined by its position in the daisy chain. Lower-priority devices may acknowledge an interrupt only when there are no higher-priority devices requesting an interrupt.

Deadlock: A condition where two parts of a program hang because they each need a resource the other has.

Debugger: A program that executes on the target system and allows the engineer to examine memory and I/O, set breakpoints, and download code, and that often supports other features. A debugger is sometimes called a *ROM monitor.*

DMA: Direct memory access. A mechanism whereby a microprocessor temporarily gives up its external bus to another processor (or other controller), and which permits the other processor to access memory directly. Some microprocessors, such as the 80186, have built-in DMA controllers.

Download: Transferring program data from a host system (such as a host PC) to a target system.

DSP: Digital signal processor. A microprocessor that is optimized for processing signals such as sound, video, or RF.

DRAM: Dynamic RAM. RAM that stores information as charge on a capacitor. Must be periodically refreshed to renew the charge and retain data.

Dynamic Memory Allocation: In a real-time kernel, the ability to dynamically allocate memory, on request, from a larger memory pool.

Dynamic Priority: In a real-time kernel, the ability to change the priority of tasks while running code.

Edge-Sensitive Interrupt: An interrupt that is recognized on a rising or falling edge.

EPROM: Erasable Programmable Read Only Memory. A PROM that can be erased using ultraviolet light.

Executive: *See* Kernel.

Firmware: Software in machine-readable form, embedded in a ROM, PROM, EPROM, Flash Memory, or other nonvolatile storage.

Flash Memory: A PROM that can be electrically erased and reprogrammed.

Fragmentation: A condition under dynamic memory allocation where memory is fragmented into many small, isolated blocks. A task that needs multiple blocks of contiguous memory may be unable to run.

Interrupt Controller: An IC or internal part of a microprocessor that prioritizes interrupts and provides a vector to the processor.

Kernel (Executive): A real-time kernel provides task management, resource management, and communication functions for a system. A kernel with file handling capability and other I/O functions is usually called a *Real-time operating system (RTOS)*.

ISR: Interrupt service routine.

Latency: The time delay between assertion of an interrupt and servicing of the interrupt.

Level-Sensitive Interrupt: An interrupt that is recognized while in the active state.

Machine Language: The binary ones and zeros that the microprocessor reads from memory and executes. *See* Assembler.

Microcontroller: A microprocessor with internal RAM and I/O ports and often including ROM, EPROM, or EEPROM for program storage.

Microprocessor: An IC containing at minimum a central processing unit (CPU), and a means to access external memory. Microprocessors may also include internal memory, I/O ports, or peripherals.

ms: millisecond. One thousandth of a second.

ns: nanosecond. One billionth of a second.

Nested Interrupts: These occur where interrupts are structured so that a lower-priority ISR can be interrupted by a higher priority ISR.

NMI: Non-maskable interrupt. An interrupt input, available on many processors, that cannot be masked off. If the interrupt occurs, the processor will always service it.

NVRAM: A package housing a static RAM IC and a battery. The battery powers the RAM so that it will retain its contents when external power is off.

OTP EPROM: One-time programmable EPROM. An EPROM without the erasure window. The OTP EPROM acts like a one-time programmable PROM, but has an EPROM structure internally.

PLD: Programmable logic device. A programmable integrated circuit used to implement logic functions.

Preemptive Scheduling: A scheduling technique where each task is given control until it finishes or is superseded by a higher priority task.

Priority Inversion: A scheduling situation where a high-priority task is forced to execute at the priority of a lower-priority task because some low-priority task has control of a resource the high-priority task needs to run.

PROM: Programmable read-only memory. A ROM that can be programmed, either by a PROM programmer or by the target system. Once programmed, acts as a read-only memory (ROM).

RAM: Random access memory. Memory that is both readable and writeable.

ROM: Read-only memory. A memory device that can be read by the central processing unit (CPU) but not written to. The contents of the ROM are fixed by a mask during manufacture.

ROM Monitor: *See* Debugger.

RTOS: Real-time operating system. Firmware that provides task scheduling, memory allocation, and other services for a real-time application.

Single Step: Causing the CPU to step through one machine or HLL instruction at a time.

Software: Computer instructions. May refer to the source code or the actual machine-readable data.

SRAM: Static RAM. RAM that is implemented as an array of flip-flops. Information is retained until overwritten or until power is removed.

Static Memory Allocation: In a real-time kernel, this refers to memory buffers and queues that are fixed at compile time and cannot be changed while running.

Static Priority: In a real-time kernel, this refers to priority that must be fixed when tasks are defined and cannot be changed at run time.

Target: The system or microprocessor that an emulator is designed to install to or replace when debugging.

UART: Universal asynchronous receiver/transmitter. An IC or circuit that provides an asynchronous serial interface.

Vector (interrupt): A number or instruction that is translated into an address, which is then executed to service an interrupt.

WDT: Watchdog timer. A timing circuit that resets or otherwise notifies a microprocessor if it is not triggered at periodic intervals.

Index